Strategic Human Resource Analytics and Intelligent Data-Driven Systems

Mufaro Dzingirai · Sulaiman Olusegun Atiku
Editors

Strategic Human Resource Analytics and Intelligent Data-Driven Systems

Prospects, Pathways and Imperatives

Springer

Editors
Mufaro Dzingirai
Harold Pupkewitz Graduate School of Business
Namibia University of Science and Technology
Windhoek, Namibia

Sulaiman Olusegun Atiku
Harold Pupkewitz Graduate School of Business
Namibia University of Science and Technology
Windhoek, Namibia

ISBN 978-981-92-2622-1 ISBN 978-981-92-2623-8 (eBook)
https://doi.org/10.1007/978-981-92-2623-8

This Springer imprint is published by the registered company Springer Nature Singapore Pte Ltd.
The registered company address is: 152 Beach Road, #21-01/04 Gateway East, Singapore 189721, Singapore

Acknowledgements

We would like to thank many people who contributed to this edited collection. We extend our deepest gratitude to the authors of all chapters contained in this volume. Their commitment to high-quality standards of research work enriched the content of this book. To our colleagues, we would like to thank them for their thought-provoking and challenging discussions on the intersectionality of digital transformation and human resource management. The scientific debates from different research philosophies and ideological orientations broadened our approach to strategic human resource analytics. We are grateful to the reviewers for their time and constructive feedback. We are also deeply indebted to the editorial and production team at Springer Nature. To our family members, we express our gratitude for their exceptional and unwavering support throughout the journey towards the completion of this edited volume.

About This Book

This book, *Strategic Human Resource Analytics and Intelligent Data-Driven Systems: Prospects, Pathways and Imperatives*, provides a comprehensive examination of how advanced and emerging technologies are reshaping human resource management processes, practices, and systems in the current digital epoch. Development of human resource analytics capabilities appears to be a powerful strategic intangible tool for outmaneuvering competitors and securing a durable competitive advantage. This necessitates a paradigm shift from viewing human resource management as purely administrative to adding value in the context of evidence-based business strategy. To ensure high return on investment, strategic human resource analytics pave the way for evidence-based decisions related to the alignment of human resource analytics with the strategic imperatives and goals of the organization. Therefore, this book delves deeply into a critical examination of the sustainable value creation, effectiveness, measurement, justification, ethical concern, and strategic impact of strategic human resource analytics and intelligence data-driven systems.

This edited volume illuminates the cross-disciplinary perspectives, multistakeholder imperatives, and multisectoral insights with the aim to enrich our understanding of the strategic value creation of human resource analytics and intelligence data-driven systems. It critically interrogates the prospects, challenges, and pathways in driving and transforming talent management discourse with the deployment of emerging technologies like Artificial Intelligence (AI), human information management systems, and machine learning. Key themes include predictive analytics, responsible human resource practices, cognitive analytics, AI threats on employee behavior, diagnostic analytics, data management, prescriptive analytics, ethical dilemmas, descriptive analytics, change management models, AI-driven talent acquisition models, and cybersecurity. Drawing on empirical studies, conceptual analysis, systematic reviews, and case studies, this edited volume presents insights into how human resource practitioners, organizations, captains of industry, and policymakers can unlock the potential value linked to digital transformation in talent management. Hence, this edited collection is of utmost importance to practitioners, scholars, policymakers, leaders of industry, software developers, and postgraduate students who are interested in harnessing the strategic value of human resource analytics and intelligence data-driven systems.

Contents

Editors and Contributors

About the Editors

Mufaro Dzingirai is a Senior Lecturer at Harold Pupkewitz Graduate School of Business, Namibia University of Science and Technology, Namibia and Research Fellow at INTI International University, Malaysia, as well as a Consultant at the Econowrite Institute for Policy Innovation & Development (EIPID), Australia.

Sulaiman Olusegun Atiku is a Director of Research and HRM Professor at Harold Pupkewitz Graduate School of Business, Namibia University of Science and Technology, Namibia and Adjunct Professor in the Department of Economic and Business Sciences, Walter Sisulu University, South Africa.

Contributors

Richard Afedzie Department of Management Studies, Pentecost University, Accra, Ghana

Sulaiman Olusegun Atiku Harold Pupkewitz Graduate School of Business, Namibia University of Science and Technology, Windhoek, Namibia; Department of Economic and Business Sciences, Walter Sisulu University, Mthatha, South Africa

Rasheed Olawale Azeez Department of Industrial Relations and Human Resource Management, Lagos State University, Ojo, Nigeria

K. A. Dhanya School of Business Management, Christ University, Bangalore, India

Mufaro Dzingirai Harold Pupkewitz Graduate School of Business, Namibia University of Science and Technology, Windhoek, Namibia;

Faculty of Business and Communications, INTI International University, Nilai, Malaysia

Neema Gupta Graphic Era Deemed to be University, Dehradun, India

Foluso Ilesanmi Jayeoba Department of Industrial Relations and Human Resource Management, Lagos State University, Ojo, Nigeria

Eugene Okyere Kwakye Department of Management Studies, Koforidua Technical University, Koforidua, Ghana

Mahadih Kyambade Department of Leadership and Governance, Makerere University Business School, Kampala, Uganda

Ibrahim Olanrewaju Lawal Department of Business Management, GBS/ University of Suffolk, Birmingham Campus, United Kingdom

Brian Mabuyana Department of Marketing and Retail Management, University of South Africa, Pretoria, South Africa

Shikha Malhotra School of Business Management, Christ University, Bangalore, India

Admore Mashokoh Department of Scholarly Communication (Research Discovery and Dissemination, Great Zimbabwe University, Masvingo, Zimbabwe

Janeth Matingwena Department of Human Resource Management, Manicaland University of Applied Sciences, Mutare, Zimbabwe

David Mhlanga Monash University, Clayton, VIC, Australia

Kebiat Mukuze Department of Human Resource Management, Midlands State University, Gweru, Zimbabwe

Afulah Namatovu Department of Applied Computing & IT, Makerere University Business School, Kampala, Uganda

Emmanuel Ndhlovu School of Public Management, Governance and Public Policy, College of Business and Economics, University of Johannesburg, Johannesburg, South Africa

Abimbola Foluso Ojapinwa Department of Industrial Relations and Human Resource Management, Lagos State University, Ojo, Nigeria

Evwiekpamare Fidelis Olori Department of Business and Entrepreneurship, GBS, Oxford Brookes University, Oxford, UK

Shikha Saloni School of Management, ITM University, Gwalior, India

H. Sandhya School of Business Management, Christ University, Bangalore, India

Ismaila Yusuf Department of Business and Entrepreneurship, Kwara State University, Malete, Nigeria

Strategic Human Resource Analytics and Intelligent Data-Driven Systems: An Introduction

Mufaro Dzingirai and Sulaiman Olusegun Atiku

Abstract There is a growing need for Human Resource (HR) practitioners to develop strategic foresight, making use of Strategic Human Resource Analytics (SHRA). Leveraging advances in innovative technologies and intelligent systems, HR practitioners can gain better insights from descriptive, diagnostic, predictive, and prescriptive intelligence using data available throughout the employee life cycle. This chapter conceptualizes and establishes the basis of integrating intelligent systems into SHRA. It demonstrates how this integration enhances HR effectiveness, supports strategic foresight, and repositions HR from an administrative support function to a proactive strategic partner. The emerging trends in SHRA are provided, highlighting its relevance for HR practitioners, and showcasing the importance of the regulatory framework for intelligence SHRA and decision-making. With such an evidence-based perspective, the chapter contributes to scholarly and practical debates on how intelligent SHRA can be leveraged to drive workforce sustainability, improve operational efficiency, and organizational outcomes in both public and private sectors.

Keywords Artificial Intelligence · Machine learning · Diagnostic analytics · Strategic human resource management · People analytics · Intelligent data-driven systems

M. Dzingirai (✉) · S. O. Atiku
Harold Pupkewitz Graduate School of Business, Namibia University of Science and Technology, Windhoek, Namibia
e-mail: mdzingirai@nust.na

S. O. Atiku
e-mail: satiku@nust.na

M. Dzingirai
Faculty of Business and Communications, INTI International University, Nilai, Malaysia

S. O. Atiku
Department of Economic and Business Sciences, Walter Sisulu University, Mthatha, South Africa

M. Dzingirai and S. O. Atiku (eds.), *Strategic Human Resource Analytics and Intelligent Data-Driven Systems*, https://doi.org/10.1007/978-981-92-2623-8_1

1 Introduction

Strategic Human Resource Analytics (SHRA) is essential for evidence-based Human Resource (HR) interventions and helps in achieving strategic goals. SHRA is a component of strategic human resource management, which is instrumental in aligning HR interventions, practices, and strategies with the business strategy. This is what makes SHRA a strategic tool for workforce planning to drive HR effectiveness. It can be described as a data-driven decision-making process that involves the use of HR metrics, collecting valid information, analyzing and interpreting results to improve decision quality and HR effectiveness, thereby contributing toward the attainment of organizational goals (Shet, 2025).

The process of collecting relevant information from clients and analyzing such information has been automated using intelligent data-driven systems (Kumari et al., 2024). The implementation of Artificial Intelligence (AI)-enabled HR systems, such as web-Based Human Resource Information Systems (HRIS) and other talent analytics platforms, assists in automating data collection throughout the employee lifecycle within an organization (Praveen & Peri, 2025; Rastogi et al., 2025). These systems also assist in conducting complex statistical analysis and providing a comprehensive interpretation of results, leading to evidence-based HR decisions (Marler & Boudreau, 2017). The integration of AI in the HR department enhances the descriptive, diagnostic, predictive, and prescriptive power of Human Resource Analytics (HRA) for strategic decision-making in the digital age (Bahuguna et al., 2024).

Adopting intelligent data-driven systems for descriptive analytics, HR practitioners can easily access historical HR data to showcase trends and patterns to assist business leaders in making informed decisions. The outputs of descriptive analytics, such as dashboards and visualizations, help in gaining insights into the overview of HR activities over a specified period. The diagnostic analytics serve as a "black box" approach, useful in understanding why and how certain events or behavioral patterns occur within a workforce. For example, diagnostic analytics provides the basis for developing specific interventions aimed at addressing issues such as high employee turnover, low commitment, and skills gaps within a workforce. Intelligent data-driven systems can perform predictive analytics using historical data to conduct statistical modeling and forecast future trends in the workforce and improve organizational outcomes. While prescriptive analytics—on the other hand—can transform those forecasts into concrete HR strategies aligned with organizational goals.

The chapter provides conceptual clarifications on SHRA and intelligent data-driven systems and establishes the basis of integrating the two concepts. This integration could be referred to as intelligent SHRA, which is a powerful tool for evidence-based Human Resource Management (HRM) practices. It shifts the HR role from administrative support to a more proactive and evidence-based decision-making department, shaping the current and future strategic direction of an organization. This chapter is important in setting the tone for the effective implementation of intelligent SHRA for evidence-based decision-making, thereby contributing to HR effectiveness in public and private sectors.

In recent times, HR effectiveness is measured by how HR strategies, interventions, policies, procedures, and practices contribute to the organizational outcomes. For HR strategies and practices to add economic value in contemporary workplaces, intelligent SHRA provides robust data for statistical modeling, leading to business intelligence that informs evidence-based decision-making. With the analytical capability of intelligent SHRA, HR leaders can design and implement strategic initiatives to ensure workforce sustainability, improve operational efficiency, and organizational outcomes. Therefore, drawing insights from the predictive and prescriptive capabilities of intelligent SHRA, HR is empowered to implement strategic initiatives that are not only aligned with the strategic goals but also contribute to business sustainability.

2 Meaning of Strategic Human Resource Analytics

SHRA represents a paradigm shift in HRM, transitioning from a traditional administrative role to a more strategic role, investing in data-driven insights to align HR strategic initiatives with long-term business goals (Nastase et al., 2025; Shet, 2025). It refers to data-driven decision-making using HR data coupled with statistical modeling to inform decision-making and drive business outcomes (Zhang, 2024). HR leaders become active strategic business partners, leading digital transformation and promoting workforce sustainability amid complex people-related business problems in the digital age through SHRA.

SHRA enables the use of HR data, such as workforce demographics, performance, engagement, and turnover, to conduct predictive and prescriptive analytics to produce results that direct the trajectory of HR initiatives (Dutraj & Sengupta, 2025). Essentially, SHRA dictates the levels of HR effectiveness in recent times. HR effectiveness in this context refers to the department's capability to transform people data into strategic insights that enable HR leaders to make informed decisions leading to sustainable competitive advantage. Most importantly, the required evidence-based strategic foresight to navigate the complex business environment in the digital age and remain competitive is dependent on robust SHRA (Donthu et al., 2024). Evidence-based strategic foresight involves making workforce predictions based on available data and results from empirical research (Al-Shammari et al., 2024). It allows the use of prescriptive analytics to create future scenarios based on information from intelligent data-driven systems for strategic decisions rather than instinctive decision-making (Atiku et al., 2024; Rahaman & Bari, 2024). An example of evidence-based strategic foresight is using predictive analytics to forecast a shortage of digital skills in the workforce in the next 3 years and developing talent pipeline strategies or new recruitment models.

3 Conceptualization of Intelligent Data-Driven Systems

An intelligent data-driven system can be defined as computer programs that handle data collection, analytics, and interpretation of empirical evidence for decision-making processes in organizations (Zong & Guan, 2025). It can be referred to as AI-enabled data analytics that turns HR data into organized and valuable information for strategic decision-making. The systems eliminate the manual process of gathering information from employees and analyzing data for informed decisions. This increases the effectiveness of HRM by reducing human errors in the decision-making process.

Integrating AI-enabled HR systems for automated data capture across the full employee lifecycle enables HR professionals to access comprehensive, real-time datasets that inform strategic interventions (Kanam, 2025). HR practitioners can automate data collection mechanisms and subsequently make evidence-based HR decisions by applying AI to HR functions and processes. Automated reporting will provide accurate and reliable HR data to monitor employees' performance or activities over a period. Instead of just performing administrative functions, HR practitioners can analyze data to gain strategic foresight in managing employees in the workplace. Smart data-driven systems build upon descriptive statistics by using statistical modeling to create descriptive, diagnostic, predictive, and prescriptive analysis (Bahuguna et al., 2024). These systems provide complex statistical analysis and compile a detailed interpretation of results for evidence-based decision-making. Descriptive analytics digests employee data to identify patterns and trends. Descriptive statistics analyze past data to identify trends. Diagnostic analytics evaluates HR data to determine why trends occurred. Predictive analytics makes use of patterns identified in data to predict future workforce needs. Prescriptive analytics, on the other hand, utilize predictions to design HR initiatives and achieve strategic goals (Kumari et al., 2024).

Dashboards, trending reports, heat maps, and risk predictions are among the business intelligence that intelligent data-driven systems can provide (Dhiman et al., 2024). The chapter suggests that descriptive analytics outputs—dashboards and data visualizations—help decision-makers understand what occurred in their HR department within a specified time. Dashboards operate at the executive level to improve strategic awareness of an organization's HR functions. By transforming infinite relational factors into digestible reports, dashboards allow HR executives to assess their workforce and predict risks (Akter & Kudapa, 2024). The chapter holds that the outputs of descriptive analytics, dashboards and visualizations, support decision-makers by offering clear insights into HR activities over a specified period, thereby enhancing strategic awareness at both operational and executive levels. These systems translate complex multivariate relationships into intuitive formats, enabling HR leaders to evaluate workforce dynamics and anticipate emerging risks with greater analytical clarity (Shafa, 2025).

Intelligent data-driven systems significantly contribute to SHRA by allowing HR leaders to base intervention programs on data. The chapter emphasizes the place of intelligent SHRA in gaining strategic foresight for data-driven interventions that support long-term organizational sustainability (Hossain et al., 2025). With this approach, the HR department can design specific interventions toward achieving sustainable competitive advantages using predictive and prescriptive analytics.

Incorporating intelligent data-driven systems for SHRA, HR practitioners are positioned as strategic business partners accountable for evidence-based strategic foresight and workforce sustainability. This comes with the right investment in intelligent systems to convert HR data into strategic initiatives to improve operational efficiency and ensure high return on investment in the digital age. HR practitioners should be empowered to make significant contributions to the organization by investing in learning and development opportunities and acquiring intelligent systems for SHRA.

4 Strategic Human Resource Analytics and Intelligent Data-Driven Systems

SHRA and intelligent data-driven systems offer structured ways of gathering and analyzing HR data to provide evidence-based strategic initiatives. Intelligent data-driven systems support SHRA by automating HR data collection and analysis processes for strategic foresight required in the digital age (Hossain et al., 2025; Zong & Guan, 2025). By implication, intelligent data-driven systems offer the information that SHRA needs to develop strategic initiatives and add economic value to the organization. The system automates SHRA processes for evidence-based HRM (Kumari et al., 2024). Intelligent data-driven systems not only provide reliable data but also reduce human errors in gathering and analyzing HR data. Therefore, using intelligent systems for SHRA drives an evidence-based decision-making process and contributes to the effectiveness of HR.

Moreover, HR practitioners are better equipped to provide actionable insights by leveraging analytical models embedded in intelligent SHRA (Mangal et al., 2024; Rahman, 2025). While applying their minds in providing evidence-based narratives to secure the buy-in of top management. This implies that AI-enabled SHRA is a technological tool that provides data-driven insights and supports the development of strategic initiatives, adding value to both employees and management (Al-Shammari et al., 2024; Bahuguna et al., 2024; Marler & Boudreau, 2017). With the analytical power in intelligent systems, HR practitioners are better equipped to provide various types of analytics that inform strategic interventions and achieve organizational outcomes (Bahuguna et al., 2024).

Linking SHRA and intelligent data-driven systems drives HR efficiency and dictates the strategic foresight required for workforce sustainability. Hence, HR practitioners are better equipped with AI-enabled analytics to predict workforce risks and provide appropriate strategic initiatives to mitigate such risks (Kumari et al., 2024). With this approach, HR takes a proactive or strategic role compared to the traditional administrative role. There is no doubt that strategic foresight is required in managing risks, which may affect workforce productivity in the short or long run. Therefore, intelligent SHRA is a powerful tool in anticipating future risks within the workforce and implementing strategic initiatives backed up with data rather than unverified assumptions. This strategic foresight strengthens the organization's preparedness for future contingencies and reinforces HR's role in long-term strategic planning (Elugbaju et al., 2024).

Furthermore, the application of intelligent SHRA shifts the focus of HRM from the administrative functions into strategic functions, leveraging the power of intelligent systems for workforce sustainability, enhanced operational efficiency, and improved organizational outcomes (Zhang, 2024). The application of intelligent SHRA in HRM enables AI-human collaborative decision-making in HR departments. This constructive collaboration brings about human-computer interactions, which are effective in addressing human errors in the decision-making process, and at the same time mitigating the accountability and ethical concerns in AI applications (Sharma et al., 2025; Tjimuku & Atiku, 2024). Therefore, intelligent SHRA is a powerful tool in enhancing decision quality and facilitating evidence-based HR interventions for operational efficiency and competitiveness in the digital age.

5 Trends in Strategic Human Resource Analytics

In the early stage, HRA initiatives were descriptive (what) and diagnostic (why) analytics but have recently evolved to predictive (what will happen) and prescriptive (what should be done) applications for proactive strategic decision-making (Levenson & Fink, 2017). According to Espegren and Hugosson (2023), many organizations operate at descriptive or diagnostic levels of HRA, which is why HR often lacks strategic foresight to effectively manage complexity in the business environment.

SHRA, on the other hand, suggests a shift from the administrative functions to strategic functions focusing on developing strategic foresight through HRA for evidence-based strategic initiatives to achieve organizational goals (Angrave et al., 2016). Recent developments in AI and other intelligent systems, such as agentic AI, have revolutionized HR operations, leading to HR effectiveness in terms of decision quality and improved workforce productivity (Al-Shammari et al., 2024). The use of agentic AI to harvest HR data on HRIS for robust predictive and prescriptive analytics can improve strategic decision-making in organizations. With this approach, intelligent systems can be incorporated in HR to harvest reliable data from HRIS and other employee engagement platforms for evidence-based HR interventions to drive workforce productivity (Al-Shammari et al., 2024). Marler and Boudreau (2017) argue that data integration challenges from multiple platforms should be addressed to maximize the full potential of intelligent systems. Hence, HR practitioners are better equipped with prescriptive analytics to implement talent development and retention strategies for workforce sustainability in the digital age.

The emerging trends in SHRA may include AI-enabled HRA, which is useful for accurate prediction of workforce risks and prescribing strategic initiatives to mitigate such risks. There is a growing support in terms of integrating intelligent systems for evidence-based decision-making to improve HR effectiveness (Căvescu & Popescu, 2025). While it is equally important to address the issues of ethical concerns in AI applications, studies suggest that sustainable application of intelligent systems requires human oversight functions to address such concerns (Angrave et al., 2016; Espegren & Hugosson, 2023; King, 2016). Judging from the foregoing, the regulatory body needs to develop a policy framework for AI-enabled HRA, particularly in public institutions operating in emerging economies. In this context, the policy framework serves as an operational guideline that supports the development of technical competencies for the responsible application of intelligent systems to develop evidence-based strategic initiatives and drive workforce productivity (Marler & Boudreau, 2017).

Current trends in intelligent SHRA involve predictive and prescriptive analytics-driven insights that help in identifying skill shortages, designing personalized development pathways aligned with future strategic needs of the organizations, and assessing the effectiveness of such strategic initiatives (Al-Shammari et al., 2024). Within SHRA, AI-enabled predictive models are currently being used to improve the accuracy of retention risk of strategically valuable employees, identify high flyers, and evaluate the long-term performance potential of employees (Dubey, 2023). The application of intelligent SHRA aligns closely with the need for workforce sustainability and achieving strategic goals in contemporary workplaces. Espegren and Hugosson (2023) establish the need for cross-cultural contextualization of SHRA, its longitudinal impacts on organizational performance, and the role of HRA in building strategic resilience amid volatility.

6 Relevance of Strategic Human Resource Analytics to Human Resources Practitioners

The integration of technological innovations in HRM over the last two decades has transformed the role of HR practitioners from administrative experts to strategic business partners. With this transition, HR practitioners are expected to develop strategic decision-making capabilities to improve HR effectiveness and make a significant contribution toward the attainment of strategic goals in contemporary workplaces (Marler & Boudreau, 2017). The use of SHRA enables HR practitioners to rely on data and advanced statistical modeling for evidence-based strategic foresight and developing initiatives to enhance HR effectiveness and improve organizational outcomes in the digital age. In the last decade, Rasmussen and Ulrich (2015) argue that organizations in which HR practitioners adopt HRA for workforce planning and talent management are more consistent in making high-quality and defensible strategic decisions. In recent times, this approach has been useful for skills gap identification, talent attraction, development, allocation, and retention to ensure workforce sustainability, operational efficiency, and sustainable competitive advantage (Atiku et al., 2024).

With SHRA, HR practitioners are empowered to identify evidence-based high-flyers, create an atmosphere where employee engagement can thrive, and ensure psychological safety within the system to promote workforce creativity and team innovation. This should be inculcated into the line managers or team leaders who have direct contact with the bottom line on a day-to-day basis. By empowering line managers, one can easily identify the early warning signs of workplace silence or quiet quitting to develop appropriate talent retention strategies (Atiku et al., 2025; Boudreau & Cascio, 2017). Accordingly, analytics-based retention strategies tend to produce better results than HR interventions based on managerial intuition, thereby establishing the strategic role of HR practitioners in managing people at work and achieving strategic goals in contemporary workplaces.

HR practitioners can enhance employee experience by generating actionable insights from employee satisfaction surveys. Working with survey data, robust data analysis, and proper interpretation of results, HR interventions can be developed to address employee dissatisfaction, thereby increasing employee satisfaction, commitment, and dedication toward the accomplishment of organizational goals (Marler & Boudreau, 2017). The foregoing could be referred to as antecedents to employee productivity. SHRA assists HR practitioners to ascertain the cause of dissatisfaction and develop evidence-based interventions to boost employee morale, create a productive work environment, and increase their levels of commitment and satisfaction, thereby improving workforce productivity (Angrave et al., 2016; Boudreau & Cascio, 2017).

Furthermore, the use of HR metrics enables line managers to objectively measure and track individual employee contributions toward departmental or organizational objectives. Using performance management systems, HR practitioners can identify the areas of strength and weakness in the workforce and develop appropriate interventions. These interventions may include incentivizing high performers, upskilling low performers, or offering negative reinforcement, thereby promoting a high-performance culture (Levenson, 2018). In this context, performance data can serve as a training needs assessment prior to the development of training programs by learning and development practitioners. While the post-training survey could be designed to measure the effectiveness of learning and development interventions. Therefore, the growing need for strategic foresight in HR functions positions HR practitioners as strategic business partners in adding economic value to the organization and achieving sustainable competitive advantage.

7 Regulatory Framework for Intelligent Strategic Human Resource Analytics

With the recent technological innovations and proliferation of intelligent systems, coupled with the need for workforce sustainability worldwide, a policy framework guiding the responsible applications of intelligent systems becomes imperative. The policy framework serves as a guideline preventing unethical use of intelligent systems and ensuring workforce sustainability. Li et al. (2023) argue that such a policy framework is fundamental in enforcing compliance concerning the integration of intelligent systems in managing people at work. Therefore, a regulatory framework for intelligent SHRA provides a policy guideline for the responsible application of data-driven systems for evidence-based workforce planning and talent retention strategies. Recent studies on responsible intelligent systems argue that AI policy should make provision for data privacy and promote human accountability (Bahangulu & Owusu-Berko, 2025; Janssen et al., 2020). In the absence of an appropriate regulatory framework, analytics-driven HR practices risk eroding organizational legitimacy, employee trust, and decision defensibility.

A regulatory framework for intelligent SHRA can be conceptualized around four interdependent dimensions, such as data governance, ethical assurance, human accountability, and strategic embeddedness (Sedeeq & Arman, 2025). The first dimension focuses on data governance, which unpacks the need for reliability, validity, and confidentiality of HR data in intelligent systems (Bena et al., 2025). This dimension is based on ensuring effective HR data governance, which involves informed written consent, protection of confidential information, and data integrity to ensure credibility of prescriptive analytics. The essence of a data governance framework is to safeguard the integrity of HR data and circumvent any litigation that may arise from the unethical use of confidential information.

The second dimension is incorporated into the framework to enforce ethical considerations in data gathering, analysis, and interpretation of results generated using intelligent systems in HR departments (Bahangulu & Owusu-Berko, 2025). Given that intelligent SHRA will rely on workforce data for prescriptive analytics, business analysts should uphold these ethical considerations in applying appropriate analytical models, decision-making processes, and providing evidence-based narrative to the top management (Sedeeq & Arman, 2025). Therefore, this dimension can be institutionalized in organizations by establishing ethics committees to ensure ethical compliance in evidence-based HRM.

Human accountability is the third dimension in the regulatory framework, which can be referred to as the cornerstone of intelligent SHRA. The reason is that HR practitioners should be at the receiving end of intelligent systems; they are responsible for the automation, accountability, and maintaining the integrity of HR decisions (Shet, 2025). Therefore, intelligent systems cannot be held accountable for errors due to poor oversight functions or system audit. Human judgment is essential for a proper interpretation of insights gained from intelligent systems, applying HR principles and human discretion before recommendations are presented to the executives

(Sedeeq & Arman, 2025). The regulatory framework establishing the place of human accountability in AI-enabled HRA re-echoes the strategic role of HR practitioners in promoting analytics-driven interventions.

The fourth dimension of the regulatory framework is fundamental in ensuring that AI-enabled HRA is strategically embedded rather than incorporated as a standalone technological intervention. This dimension focuses on integrating smart systems with the strategic priorities of an organization. In this context, emphasis could be placed on how intelligent systems will equip HR practitioners with strategic foresight to improve workforce productivity, add economic value to the organization, and achieve sustainable competitive advantage (Tuli et al., 2018). These four regulatory dimensions provide robust policy guidelines for the effective integration of intelligent systems in SHRA to increase HR effectiveness and make a significant contribution to organizational competitiveness in the digital age.

This book is organized as follows: Chapter: "Strategic Human Resource Analytics and Intelligent Data-Driven Systems: An Introduction". This chapter conceptualizes and establishes the basis of integrating intelligent systems into SHRA. It demonstrates how this integration enhances HR effectiveness, supports strategic foresight, and repositions HR from an administrative support function to a proactive strategic partner. The emerging trends in SHRA are provided, highlighting its relevance for HR practitioners, and showcasing the importance of the regulatory framework for intelligence SHRA and decision-making. With such an evidence-based perspective, the chapter contributes to scholarly and practical debates on how intelligent SHRA can be leveraged to drive workforce sustainability, improve operational efficiency, and organizational outcomes in both public and private sectors.

Chapter: "The Role of Strategic Human Resources Analytics in Promoting Organizational Performance: A Case Study of State Universities in Zimbabwe". Using the mixed method approach, this chapter examines the role of strategic human resources analytics in promoting organizational performance in three identified state universities. The research study revealed that Zimbabwean state institutions of higher learning lag in strategic human resource analytics adoption due to multiple challenges like financial restrictions, skill gaps, and cultural opposition. Nevertheless, participants acknowledge the possibility of improving productivity, equity, and important decision-making. A government-backed model, combined with training and low-cost technology, could drive SHRA integration. Future research should explore cost-effective analytics tools and change-management strategies for higher education. The findings contribute to the growing body of strategic human resource analytics and provide a unique contribution to theory, practice, and policy in higher education.

Chapter: "The Influence of AI Threats on Employee Behavior and Turnover Intentions". This chapter explores the relationship between the Threat of Artificial Intelligence (TAI) and employee turnover intention, emphasizing the roles of role ambiguity and knowledge hiding behavior. This study employs a quantitative approach using PLS-SEM to analyze how perceived Artificial Intelligence (AI) threats influence turnover intentions. The findings indicated that the Threat of AI (TAI) had a significant positive impact on Turnover Intention ($\beta = 0.289, p < 0.001$), Role Ambiguity ($\beta = 0.368, p < 0.001$), and Knowledge Hiding Behavior ($\beta = 0.670$,

$p < 0.001$). Additionally, Role Ambiguity ($\beta = 0.119$, $p < 0.01$) and Knowledge Hiding Behavior ($\beta = 0.259$, $p < 0.01$) significantly increased turnover intention, while the mediating effect of role ambiguity (H4) was not supported. The implications of these findings highlight the need for organizations to foster transparent communication and supportive environments to reduce role ambiguity and prevent knowledge hiding.

Chapter: "Enhancing Employee Data Management in Uganda's Private Sector: Challenges, Benefits, and Pathways to Responsible Practices". This chapter combines a literature review with empirical evidence gathered via case studies, semi-structured interviews, and surveys, which is why it is a mixed-method study. Thematic analysis showed that although a few organizations have implemented organized data protection, some organizations have issues with compliance, technological constraints, and resource constraints. The results provide some of the major advantages of responsible data management, such as more trust between the employer and the employees, better legal and regulatory adherence, and more operational efficiency. Nevertheless, there are still major hindrances that include a lack of awareness, a poor application of data protection regulations, and the high cost of installing an effective data protection system. The chapter highlights the importance of well-defined policies, educating employees, and enhanced regulatory control to guarantee that the use of employee data in the Ugandan privacy sector is carried out in an ethical and efficient way.

Chapter: " **Leveraging Personalized Employee Development Plans to Support Change Management Models in Academic Libraries in Zimbabwe**". This chapter explores how personalized employee development plans assist academic libraries in Zimbabwe in implementing change management models. The study found that most academic libraries did not have custom employee development plans to support their change process. The study further stated that the potential value of custom employee development plans could support academic libraries in training their librarians with the skills needed to enable them to manage change. The study suggested that the personalized employee development plans supported strategic planning, communication, and collaboration. The study concluded that the academic libraries in Zimbabwe would benefit from using custom employee development plans to support change management models. The study recommended that academic libraries should embrace custom staff development plans to enhance librarians' capacity for managing change and embed them into their change management component.

Chapter: " Exploration of Artificial Intelligence as a Catalyst for Human Resource Management Effectiveness". This research investigated the relationship between predictive performance analytics and performance management. It also examined the effect of the robotics automation process on the recruitment and selection process. Descriptive research design was adopted in this study using 148 workers of Beacongate Limited as the study population. Research findings confirmed a correlation between the utilization of predictive performance analytics and performance management and likewise an influence of robotics process automation on the recruitment and selection process. This study offered insightful information about how to integrate AI technology into an organization's human resource management strategy for effectiveness.

Chapter: " Ethics In Human Resource Analytics: Perspectives from Public and Private Organizations in Ghana". This chapter examines the ethical ramifications of Human Resource Analytics (HRA) within Ghana's public and private sectors, focusing on justice, privacy, transparency, and ethical governance. Five major themes emerged in relation to problems, namely, algorithmic bias and fairness, data privacy and surveillance, employee autonomy and permission, profiling and behavioral trust, and organizational culture and ethical governance. The results showed that in Ghanaian businesses, the lack of digital infrastructure, low levels of data literacy, and weak ethical frameworks made discrimination and privacy violations more likely. Cultural hierarchy and regulatory gaps often make it harder to be open and responsible when using analytics. However, some companies that are ahead of the curve, especially in banking and telecommunications, have started to include fairness audits, ethical review processes, and compliance with Ghana's Data Protection Act (2012, Act 843). The study contributes to the evolving African discourse on ethical analytics by illustrating that ethical governance, trust, and cultural alignment are essential for responsible data practices.

Chapter: " Adaptive Talent Acquisition Models for Navigating Change in a Dynamic Environment". This chapter examines the transformation of talent acquisition models for contributing toward organizational success in a dynamic environment. In a country like India, labor markets are dynamic, and skill set requirements are also ever-changing. Hence, the initial part of the study examines the features of various types of talent acquisition models followed in India and also at the global level. As a progress study exhibits the talent acquisition process with special emphasis on employer branding and Employee Value Proposition (EVP). The study also delves into the impact of technology and automation on talent acquisition practices. The last part of the study examines the challenges of AI in HR practices, especially in the context of the Volatile, Uncertain, Complex, and Ambiguous (VUCA) world and concludes by sketching the road ahead for talent acquisition models, which leverage data-driven insights, technology, and agile recruitment strategies to build resilient talent pipelines that align with evolving business needs.

Chapter: " Cyber Security Vulnerability and Organizational Performance: A Systematic Literature Review". This chapter interrogates the cybersecurity weaknesses and how this weakness affects the organizational efficiency. To examine the associations between cybersecurity threats, such as those of hacktivism, cybercrime, and cyber spying and the key performance indicators, such as reputational damage, operational effectiveness, and financial stability, a systematic literature review method was used. Data were collected via EBSCOhost, Elsevier, ProQuest, Wiley-20, SpringerLink, and Google Scholar, and were reviewed and analyzed through Systematic Reviews and Meta-analyses (PRISMA). As technology and information systems expand globally, businesses must stay ahead of the curve by enhancing security measures and proactively resolving authentication and authorization vulnerabilities that could jeopardize confidentiality, and availability. Research shows that hacktivism, cybercrime, and cyber espionage harm organizational performance. To combat these risks, companies must create backup systems to protect client and customer information from hostile threats.

Chapter: " Ethical Dilemmas of Generative AI and Large Language Models in Human Resource Management". The rapid technological advancements witnessed in the early twenty-first century are reforming human resource management at a faster pace. However, there is scant evidence pertaining to the influence of generative AI and Large LLMs on human resource management. This chapter seeks to bridge this gap. This chapter aims to examine how generative AI and LLMs influence talent management in a data-oriented context. A desktop approach and conceptual approach were used as the data collection technique. The results revealed that making use of generative AI and LLMs alters the human resource management within a data-oriented setting. On the other side, the results showed that risks and ethical dilemmas affect the use of generative AI and LLMs within the data-oriented workplace. These results are useful for human resource practitioners, policy makers, and scholars.

Chapter: "Human Resource Analytics: Challenges and Prospects for Human Resource Effectiveness". Although Human Resource Analytics (HRA) is increasingly used, and the question of how it can influence employee productivity remains unanswered. Human Resource Analytics is an enhancement of Human Resources practice that uses data analysis to advance HR practices, improve employee engagement, performance, and decision-making, and enhance organizational growth, efficiency, and strategic workforce planning. The current research aims to fill this gap by investigating the impact of HRA on job satisfaction, engagement, performance, and other aspects of the workplace. The study adopted the Systematic Reviews and Meta-analyses (PRISMA) to achieve smooth, superior, and improved comprehension of the examination of the gathered information and analysis. The results indicate that HRA is positively related to employee satisfaction, engagement, and overall performance, thereby enhancing HR's capacity to address workforce needs. Nevertheless, issues of low data quality, lack of analytical skills, and ethical concerns, especially regarding privacy and fairness, affect the effectiveness and credibility of HRA initiatives. To get the best out of HRA, the chapter suggests that organizations adopt effective data management systems, improve HR professionals' analytical capabilities through training, and develop ethical standards. Such measures will contribute to transparency, employee confidence, and the successful implementation of data-oriented HR practices.

Chapter: "Strategic Human Resource Analytics and Intelligent Data-Driven Systems: A Conclusion". This concluding chapter interrogates the drivers of the adoption of strategic human resource analytics and data-driven systems, challenges faced in strategic human resource analytics, in the contemporary digital economy, practical and policy implications, and the further research agenda. This chapter concludes that the strategic human resource analytics and data-driven systems are the heart and soul of evidence-based strategic decisions for sustainable value creation through the robust application of descriptive, diagnostic, predictive, and cognitive analytics.

8 Conclusion

This chapter examined the evolving role of SHRA and intelligent data-driven systems in reshaping the HR function in the digital age. SHRA and intelligent data-driven systems offer structured ways of gathering and analyzing HR data to provide evidence-based strategic initiatives. Intelligent data-driven systems support SHRA by automating HR data collection and analysis processes for strategic foresight required in the digital age. Therefore, linking SHRA and intelligent data-driven systems drives HR efficiency and dictates the strategic foresight required for workforce sustainability. The current trends in intelligence SHRA involve predictive and prescriptive analytics-driven insights that help in identifying skill shortages, designing personalized development pathways aligned with future strategic needs of the organizations, and assessing the effectiveness of such strategic initiatives. Within SHRA, AI-enabled predictive models are currently being used to improve the accuracy of retention risk of strategically valuable employees, identify high flyers, and evaluate the long-term performance potential of employees. This shift positions HR as a strategic partner in complex and data-intensive business environments. Technological innovations in HRM over the last two decades have transformed the role of HR practitioners from administrative experts to strategic business partners. With this transition, HR practitioners are expected to develop strategic decision-making capabilities to improve HR effectiveness and make a significant contribution toward the attainment of strategic goals in contemporary workplaces. However, this chapter holds that AI policy should be developed to make provision for data privacy and promote human accountability for the responsible application of intelligent systems in HR departments.

References

Akter, M., & Kudapa, S. P. (2024). A comparative analysis of artificial intelligence-integrated bi dashboards for real-time decision support in operations. *International Journal of Scientific Interdisciplinary Research, 5*(2), 158–191.

Al-Shammari, M., Ahmed Al Bin Ali, F., Abdulla AlRashidi, M., & Salem Albuainain, M. (2024). Big data and predictive analytics for strategic human resource management: A systematic literature review. *International Journal of Computing and Digital Systems, 17*(1), 1–9.

Angrave, D., Charlwood, A., Kirkpatrick, I., Lawrence, M., & Stuart, M. (2016). HR and analytics: Why HR is set to fail the big data challenge. *Human Resource Management Journal, 26*(1), 1–11.

Atiku, S. O., Jeremiah, A., & Genty, K. I. (2025). Effects of workplace silence, quiet quitting, and quiet firing on employee turnover: Evidence from the service industry. *International Journal of Organizational Analysis*. https://doi.org/10.1108/IJOA-04-2025-5389

Atiku, S. O., Menjengua, D. K., Jeremiah, A., & Villet, H. J. (2024). Data-driven talent management practices for eco-innovation in state-owned enterprises. In S. Singh, S. Rajest, S. Hadoussa, A. Obaid, & R. Regin (Eds.), *Data-driven decision making for long-term business success* (pp. 369–394). IGI Global Scientific Publishing. https://doi.org/10.4018/979-8-3693-2193-5.ch025

Bahangulu, J. K., & Owusu-Berko, L. (2025). Algorithmic bias, data ethics, and governance: Ensuring fairness, transparency, and compliance in AI-powered business analytics applications. *World Journal of Advanced Research and Reviews, 25*(2), 1746–1763.

Bahuguna, P. C., Srivastava, R., & Tiwari, S. (2024). Human resources analytics: Where do we go from here? *Benchmarking: An International Journal, 31*(2), 640–668.

Bena, Y. A., Ibrahim, R., Mahmood, J., Al-Dhaqm, A., Alshammari, A., Nasser, M., et al. (2025). Big data governance challenges arising from data generated by intelligent systems technologies: A systematic literature review. *IEEE Access, 13*, 12859–12888.

Boudreau, J., & Cascio, W. (2017). Human capital analytics: Why are we not there? *Journal of Organizational Effectiveness: People and Performance, 4*(2), 119–126.

Boudreau, J. W., & Ramstad, P. M. (2007). Beyond HR, the new science of human capital: Harvard Business School press. *Boston, MA*.

Căvescu, A. M., & Popescu, N. (2025). Predictive analytics in human resources management: Evaluating AI's role in talent retention. *AppliedMath, 5*(3), 99.

Davenport, T. H., Harris, J., & Shapiro, J. (2010). Competing on talent analytics. *Harvard Business Review, 88*(10), 52–58.

Dhiman, R., Miteff, S., Wang, Y., Ma, S. C., Amirikas, R., & Fabian, B. (2024). Artificial intelligence and sustainability—A review. *Analytics, 3*(1), 140–164.

Donthu, S., Acharya, B., Hassan, M., Prasad, S., & Mahapatro, S. K. (2024). HR analytics: Leveraging big data to drive strategic decision-making in human resource management. *Journal of informatics education and research, 4*(1), 1541–1549.

Dubey, N. (2023). A review of literature on the use of HR analytics in decision-making. *International Journal for Research Trends and Innovation, 8*(10), 45–58.

Dutraj, R., & Sengupta, P. R. (2025). Aligning human resources to businesses through human resource analytics. *Asian Journal of Management, 16*(2), 73–81.

Elugbaju, W. K., Okeke, N. I., & Alabi, O. A. (2024). Human resource analytics as a strategic tool for workforce planning and succession management. *International Journal of Engineering Research and Development, 20*(11), 744–756.

Espegren, Y., & Hugosson, M. (2023). HR analytics-as-practice: A systematic literature review. *Journal of Organizational Effectiveness: People and Performance, 12*(5), 83–111.

Hossain, Q., Ikbal, M. Z., & Rahman, M. M. (2025). A metadata-driven decision support in human capital management: Reviewing HRIS and predictive analytics integration. *ASRC Procedia: Global Perspectives in Science and Scholarship, 1*(01), 215–246.

Janssen, M., Brous, P., Estevez, E., Barbosa, L. S., & Janowski, T. (2020). Data governance: Organizing data for trustworthy artificial intelligence. *Government Information Quarterly, 37*(3), 101493.

Kanam, S. (2025). AI and machine learning for predictive workforce analytics: Transforming data into actionable insights for real-time decision-making. In *International conference on sustainable development through machine learning, AI, and IoT* (pp. 280–295). Cham.

King, K. G. (2016). Data analytics in human resources: A case study and critical review. *Human Resource Development Review, 15*(4), 487–495.

Kumari, G. V., Chapa, B. P., Chaitanya, N. K., Mahajan, R. G., & Shahakar, M. (2024). Introduction to data-driven intelligent systems. In *Data-driven systems and intelligent applications* (pp. 1–18). CRC Press.

Levenson, A. (2018). Using workforce analytics to improve strategy execution. *Human Resource Management, 57*(3), 685–700.

Levenson, A., & Fink, A. (2017). Human capital analytics: Too much data and analysis, not enough models, and business insight. *Journal of Organizational Effectiveness: People and Performance, 4*(2), 145–156.

Li, W., Yigitcanlar, T., Browne, W., & Nili, A. (2023). The making of responsible innovation and technology: An overview and framework. *Smart Cities, 6*(4), 1996–2034.

Mangal, U., Mogha, S., & Malik, S. (2024, September). Data-driven decision making: Maximizing insights through business intelligence, artificial intelligence and big data analytics. In *In*

2024 international conference on advances in computing research on science, engineering and technology (ACROSET) (pp. 1–7). IEEE.
Marler, J. H., & Boudreau, J. W. (2017). An evidence-based review of HR analytics. *The International Journal of Human Resource Management, 28*(1), 3–26.
Nastase, C., Adomnitei, A., & Apetri, A. (2025). Strategic human resource management in the digital era: Technology, transformation, and sustainable advantage. *Merits, 5*(4), 23.
Praveen, R. V. S., & Peri, S. S. S. R. G. (2025). Smart HR. In *Leveraging AI & ML for human resource transformation*. Addition Publishing House.
Rahaman, M. A., & Bari, M. D. (2024). Predictive analytics for strategic workforce planning: A cross-industry perspective from energy and telecommunications. *International Journal of Business Diplomacy and Economy, 3*(2), 14–25.
Rahman, M. M. (2025). Data analytics for strategic business development: A systematic review analyzing its role in informing decisions, optimizing processes, and driving growth. *Journal of Sustainable Development and Policy, 1*(01), 285–314.
Rasmussen, T., & Ulrich, D. (2015). Learning from practice: How HR analytics avoids being a management fad. *Organizational Dynamics, 44*(3), 236–242. https://doi.org/10.1016/j.orgdyn.2015.05.008
Rastogi, P., Khandelwal, S., Khurana, R., Gupta, H., & Singh, P. (2025). *Revolutionizing HR: The AI-powered guide to modern HR practices*. Academic Enclave.
Sedeeq, F. S., & Arman, P. (2025). Human-centric AI governance: An adaptive public international law framework for ethical and inclusive AI regulation in public health. *Journal of Law, Medicine & Ethics*, 1–12.
Shafa, H. (2025). Artificial intelligence-driven business intelligence models for enhancing decision-making in US enterprises. *ASRC Procedia: Global Perspectives in Science and Scholarship, 1*(01), 771–800.
Sharma, H., Ali, A., Singh, M., Singhal, A., Ghai, S., & Saluja, K. (2025). Big data and artificial intelligence for strategic human resource management. In H. Rahiman & R. Kodikal (Eds.), *Transforming organizational culture through meta-driven human resources* (pp. 405–426). IGI Global Scientific Publishing. https://doi.org/10.4018/979-8-3373-0720-6.ch014
Shet, S. V. (2025). Developing human resource analytics capability architecture in organizations. *Personnel Review, 54*(9), 2286–2310.
Tjimuku, M., & Atiku, S. O. (2024). Integrating digital and human labor for sustainable production. In S. Atiku, A. Jeremiah, E. Semente, & F. Boateng (Eds.), *Eco-innovation and sustainable development in industry 5.0* (pp. 267–289). IGI Global. https://doi.org/10.4018/979-8-3693-2219-2.ch013
Tuli, F. A., Varghese, A., & Ande, J. R. P. K. (2018). Data-driven decision making: A framework for integrating workforce analytics and predictive HR metrics in digitalized environments. *Global Disclosure of Economics and Business, 7*(2), 109–122.
Zhang, L. (2024). Data-driven decision support framework for human resource strategic planning. *International Journal of High-Speed Electronics and Systems, 2540197.*
Zong, Z., & Guan, Y. (2025). AI-driven intelligent data analytics and predictive analysis in industry 4.0: Transforming knowledge, innovation, and efficiency. *Journal of the Knowledge Economy, 16*(1), 864–903.

Mufaro Dzingirai is a Senior Lecturer at Harold Pupkewitz Graduate School of Business, Namibia University of Science and Technology, Namibia and Research Fellow at INTI International University, Malaysia, as well as a Consultant at the Econowrite Institute for Policy Innovation & Development (EIPID), Australia.

Sulaiman Olusegun Atiku is a Director of Research and HRM Professor at Harold Pupkewitz Graduate School of Business, Namibia University of Science and Technology, Namibia and Adjunct Professor in the Department of Economic and Business Sciences, Walter Sisulu University, South Africa.

The Role of Strategic Human Resources Analytics in Promoting Organizational Performance: A Case Study of State Universities in Zimbabwe

Kebiat Mukuze

Abstract Strategic human resource analytics play a significant role in promoting organizational performance. Using the mixed-methods approach, this chapter examines the role of strategic human resources analytics in promoting organizational performance in three identified state universities. Organizational performance and human capital theories form the theoretical framework for this study. Closed-ended questions and a semi-structured interview guide were used to collect data for this study. A sample size of 216 respondents obtained through stratified random sampling and purposive sampling methods was used for the study. Pearson product-moment correlation analysis techniques and deductive thematic data analysis techniques were used to analyze the obtained research data. The research study revealed that Zimbabwean state institutions of higher learning lag in strategic human resource analytics adoption due to multiple challenges like financial restrictions, skill gaps, and cultural opposition. Nevertheless, participants acknowledge the possibility of improving productivity, equity, and important decision-making. A government-backed model, combined with training and low-cost technology, could drive SHRA integration. Future research should explore cost-effective analytics tools and change-management strategies for higher education. The findings contribute to the growing body of strategic human resource analytics and provide a unique contribution to theory, practice, and policy in higher education.

Keywords Data-driven decision-making · Human Capital · Organizational performance · Strategic human resource analytics · State universities · Zimbabwe

---

K. Mukuze (✉)
Department of Human Resource Management, Midlands State University, Gweru, Zimbabwe
e-mail: mukuzek@staff.msu.ac.zw

M. Dzingirai and S. O. Atiku (eds.), *Strategic Human Resource Analytics and Intelligent Data-Driven Systems*, https://doi.org/10.1007/978-981-92-2623-8_2

1 Introduction

Business organizations and academia are increasingly adopting data-driven approaches to decision-making with the intention of improving work performance. Thus, Strategic Human Resource Analytics (SHRA) has emerged as a critical tool in optimizing workforce productivity, talent management, and organizational effectiveness (Bassil, 2011). Higher Education Institutions (HEIs), particularly state universities in Zimbabwe, face challenges such as lack of funding, high staff turnover, and performance management, necessitating the adoption of SHRA to improve institutional outcomes (Marongedza et al., 2023).

Despite the growing recognition of SHRA in corporate settings, its application in Zimbabwean state universities remains underexplored. This study sought to bridge this gap by examining how SHRA contributes to organizational performance in three selected state universities in Zimbabwe. State universities in Zimbabwe operate in a dynamic and resource-constrained environment, requiring efficient Human Resource (HR) strategies to maintain competitiveness. Public universities in Zimbabwe are governed by one parent ministry of higher and tertiary education, and one remuneration grading system. Thus, the researcher opined that a sample of just three state universities would enable generalization of results to the larger population since the population was homogeneous (Schindler, 2023). However, the extent to which these institutions utilize SHRA to enhance performance remains unclear. Existing studies (Muchowe & Mubango, 2024; Ngwendere & Mlitwa, 2025) on HR analytics in Zimbabwe primarily focused on private-sector organizations, leaving a research gap in the public higher education sector. This study investigates the role of SHRA in improving organizational performance, identifying key metrics, challenges, and potential frameworks for implementation. Therefore, the specific research objectives guiding the study were to:

(1) examine the level of strategic human resource analytics employed in state universities,
(2) identify the essential human resources metrics utilized to enhance organizational performance in higher education institutions,
(3) establish the influence of strategic human resources on employee productivity in state universities,
(4) analyze obstacles to implementing strategic human resource analytics in Zimbabwean state universities,
(5) suggest a framework for incorporating strategic human resource management into higher education practices to enhance organizational performance.

The study would bring a new contextual framework for HRA in Zimbabwean higher education institutions and provide empirical data for HRA's influence on business performance in Zimbabwe and policy recommendations for the Ministry of Higher and Tertiary Education to support HR analytics implementation.

2 Literature Review

In this subsection, the researcher sought to present the current state of knowledge on the subject matter of strategic human resource analytics and organizational performance in organizations. This was possible through a scholarly survey of similar context studies to gain familiarity with the topic and its scholarly context, to position the current work in relation to other researchers and theorists (McCombes, 2023). Information in this section is presented following the order of theoretical framework supporting the study, empirical evidence literature review, and hypotheses development (Creswell, 2014).

2.1 Theoretical Framework

The book chapter was anchored on Organizational Performance Theory and Human Capital Theory, furnishing a robust foundation for understanding the role of Strategic Human Resource Analytics (SHRA) in improving organizational operations in Zimbabwean state universities.

2.1.1 Organizational Performance Theory

The organizational performance theory clarifies that multiple performance perspectives are drawn from Peter Principle, Drucker's management by objectives (MBO) and systems thinking. The Peter Principle posits that employees in hierarchical organizations tend to rise to their level of incompetence, leading to inefficiencies (Peter & Hull, 1969). This theory suggests that without proper performance analytics, organizations risk promoting employees beyond their optimal capability, negatively affecting productivity (Peter & Hull, 1969). Similarly, Drucker's (1954) MBO emphasizes aligning employee performance with organizational goals through measurable outcomes (Drucker, 1954). In the context of Zimbabwean universities, where bureaucratic structures dominate, SHRA can mitigate these inefficiencies by identifying high-potential employees through data-driven talent assessments, detecting skill gaps, recommending targeted training interventions, and aligning HR strategies with institutional objectives (Ackoff, 1971). Strategic human resource analytics can be the modern times engine that powers the MBO through defining, tracking, and analyzing measurable outcomes. In the Zimbabwean context, where state universities are characterized by bureaucratic business practices, strategic human resource analytics can institutionalize management by objectives by using performance data to set quantifiable objectives for university staff to monitor progress through performance indicators and align individual employee contributions with organizational goals.

2.1.2 Human Capital Theory

Becker's (1964) Human Capital Theory asserts that employee skills, knowledge, and abilities are critical assets that drive organizational success. Investments in workforce development (for example, training, analytics-driven talent management) yield long-term productivity gains (Becker, 1964). In higher education, SHRA facilitates predictive analytics for faculty retention and succession planning, Return on Investment (ROI) measurement of training programs (Lengnick-Hall et al., 2018), and strategic workforce planning to align human capital with institutional needs. Strategic human resource analytics provides an important link between human capital investment and tangible organizational outcomes by quantifying the economic value and effect of capital. For example, Human Capital Theory maintains investing in top talent. Strategic human resource analytics operationalizes this by using predictive models to identify members of the faculty who can leave the organization soon and allows the university to make immediate plans, which include professional development and targeted retention packages informed by data to protect its human capital. The Human Capital Theory advocates investment in training and strategic human resource analytics formalizes this investment by measuring return on investment of the training programs, analyzing performance like post-training performance improvement and research output or learners' satisfaction scores.

These theories collectively justify the need for data-driven HR decision-making in Zimbabwean universities, where resource constraints and performance inefficiencies persist (Zishiri, 2025). Strategic human resource analytics emerge as an essential tool and mechanism that applies the aforesaid theoretical principles to solve haunting problems like inefficiencies and resource constraints.

2.2 Conceptual Framework

A conceptual framework provides a visual or narrative representation of the key variables, relationships, and theoretical foundations guiding a study (Ravitch & Riggan, 2016). For this research, the framework integrates Organizational Performance Theory (Drucker, 1954; Peter & Hull, 1969) and Human Capital Theory (Becker, 1964) to explain how Strategic Human Resource Analytics (SHRA) influences organizational performance in Zimbabwean state universities. The framework adopts both the narrative and visual representation of the conceptual framework, and is structured around inputs (SHRA adoption, HR metrics, and organizational factors), mediating processes (employee productivity, decision-making quality), outcomes (institutional performance), and contextual barriers (funding, resistance, skill gaps).

2.2.1 Components of the Conceptual Framework

Input Variables

(a) **Strategic HR Analytics (SHRA) Adoption**

Strategic human resource analytics in this study is defined as the systematic use of workforce data to inform HR decisions (Bassil, 2011). Fundamental metrics used in the practice of human resource management in different organizations include talent acquisition analytics (for example, time-to-hire and quality-of-hire). Other metrics also include retention and turnover analytics which capture attrition rates and organizational exit interview insights. Performance analytics include competency assessments and training rate of Return on Investment (ROI). SHRA enables evidence-based HR strategies, aligning workforce planning with institutional goals (Kiran et al., 2024).

(b) **Organizational Enablers**

There are various human resource analytics adoption enablers in the form of leadership support and commitment from university management (Mukuze, 2023). Technological infrastructure, like Human Resource Information Systems (HRIS) and analytics tools (Mapuranga et al., 2024), assists in the implementation of human resource analytics. HR competencies in the form of data literacy among HR professionals (Bahuguna et al., 2023) are another enabler of human resource analytics adoption.

Mediating Processes

(c) **Data-Driven Decision-Making**

SHRA converts raw HR data into actionable insights, improving recruitment efficiency (reducing bias in hiring), training effectiveness (identifying skill gaps), and succession planning (predicting leadership needs). Research established that organizations using HR analytics report 19% higher profitability (Bersin, 2023).

(d) **Employee Productivity and Engagement**

Employee productivity and engagement can be improved through different mechanisms which include personalized learning and development (Lengnick-Hall et al., 2018) and performance feedback loops (Diefenbach et al., 2024). Research findings revealed that engaged employees exhibit 21% higher productivity (Gallup, 2023).

Outcome Variable: Organizational Performance

Organizational performance in service-providing organizations like state universities can be measured through academic output like research publications and graduate

Table 1 Contextual barriers (moderating variables)

Barrier	Impact on SHRA Adoption	Supporting Evidence
Funding constraints	Limits analytics software/training budgets	Mukuze (2023)—Zimbabwean universities face austerity
Resistance to change	Cultural inertia slows adoption	Sixpence et al. (2023)—Mismatch in HR/business goals
Skill gaps	HR staff lack data analysis expertise	Sungwa (2025)—African SMEs face similar issues

Source Literature Review (2025)

employability. Performance in educational institutions can also be measured through operational efficiency in the form of staff-to-student ratios and administrative costs, among others. Stakeholder satisfaction (student feedback, accreditation ratings) is also a good measure of organizational performance in service-providing organizations. SHRA optimizes resource allocation, and this is a critical need in Zimbabwe's underfunded universities (Zishiri, 2025).

The indicated barriers in Table 1 are found in other African countries, but they are more pronounced in Zimbabwe due to various contributing factors like prolonged hyperinflation, brain drain, which created a local data analytics skills gap and austerity measures. The brain drain challenge created a local data analytics skills gap, which resulted in a challenging environment for strategic human resource analytics adoption. Thus, employees who lack data analytics skills are likely to feel threatened by new data systems and eventually resist change by sticking to old ways of human resources reporting. Widespread organizational resistance to change will undermine the business case for strategic human resource analytics and will result in leadership becoming hesitant to allocate scarce funding to human resource analytics initiatives.

2.2.2 Visual Representation of the Framework

The conceptual framework, as presented in Fig. 1, for this study is justified as it is aligned with the theoretical framework. Human Capital Theory (Becker, 1964) operationalizes Becker's (1964) premise that investments in workforce analytics yield productivity gains. While Organizational Performance Theory (Drucker, 1954) aligns with Drucker's MBO by using metrics to track goal attainment. Empirical support of the conceptual framework is got from global evidence. Past research findings established that organizations using HR analytics achieve 8% higher revenue per employee (McKinsey, 2024). In the African context, Boakye and Lamptey (2020) found that HR analytics improves retention in Ghanaian firms. From the framework above, a Zimbabwean gap exists where no prior framework exists for state universities (Mukuze, 2023). In the circumstances, the conceptual framework has practical utility for policymakers. It guides the Ministry of Higher Education in funding HR

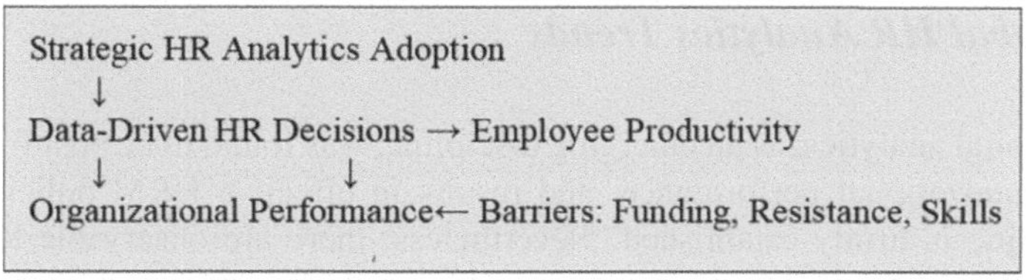

Fig. 1 Conceptual framework linking SHRA to performance in Zimbabwean universities. *Source* Researcher's Own Drawing (2025)

technology upgrades. As for universities, it helps prioritize metrics such as retention over recruitment.

Concisely, this framework bridges theory (Human Capital and Performance Theories) and empirical evidence (Global/African studies) to propose a Zimbabwe-specific SHRA model. It addresses how SHRA improves performance (via data-driven decisions), explains why adoption lags (barriers like funding), and identifies the interventions that work for SHRA adoption (training, leadership buy-in).

2.3 Empirical Evidence Literature Review

The empirical literature on Strategic Human Resource Analytics (SHRA) reveals a growing body of evidence demonstrating its impact on organizational performance across various sectors and geographical contexts. While SHRA has gained significant traction in corporate environments (Bassil, 2011), its application in Higher Education Institutions (HEIs), particularly in resource-constrained settings like Zimbabwe, remains under-explored (Mukuze, 2023). This section synthesizes empirical studies from global trends, African contexts, and Zimbabwe-specific research to establish the proven benefits of SHRA in enhancing workforce productivity and decision-making, identify key challenges hindering its adoption in different organizational settings, and identify critical research gaps in Zimbabwean state universities that justify this study.

Following Webster and Watson's (2002) recommendation for systematic literature reviews, this analysis adopts a three-tiered approach of global HR analytics trends by establishing SHRA's universal applicability, adopting African contexts, highlighting regional adoption patterns and barriers and Zimbabwe-Specific Studies, to identify local gaps in HEIs. This structure ensures a progressive narrowing from broad international evidence to context-specific insights, aligning with Ravitch and Riggan's (2016) conceptual framework development methodology. This review thus sets the stage for investigating how SHRA can be optimized in Zimbabwe's state universities, addressing both theoretical and practical gaps.

2.4 Global HR Analytics Trends

Human capital analytics, as an emerging discipline, was found to be positively influencing organizational performance, and results in effective HRM policies where the discipline is firmly established. Nevertheless, there are observable leadership challenges, lack of appropriate analytical skills and data security issues in the effective implementation of strategic human resource analytics in organizations (Kiran et al., 2024). Prima et al. (2024) study results revealed that human capital analytics moderate employee learning and development, improving organizational knowledge sharing and creativity. The same point was supported by Nowicka et al. (2024) and Diefenhardt et al. (2025) research studies, which disclosed that judicious use of data analytics in people management improves personnel matters and assists in the achievement of strategic goals of the organization. However, the issues of discrimination, data storage, and transmission security were noted as some obstacles to the full implementation of strategic human resources analytics in organizations. Al-Shammari et al.'s (2025) study results disclosed that poor data quality hinders the full utilization of strategic human resource analytics in organizations across the globe. Another study by Bahuguna et al. (2023) discovered that HRA is still a novel management science but organizations in Canada, the United Kingdom, the United States of America (USA), China, and India are slowly embracing the subject matter and its associated benefits of reducing recruitment costs and enabling the achievement of organizational goals.

2.5 African Contexts

Sungwa's (2025) research study in South African Small to Medium Enterprises (SMEs) revealed that organizations under study lack the services of properly qualified human resources professionals and the use of technology, which are all essential requirements for the adoption of human resource analytics. Tessema et al.'s (2025) research results in Ethiopia revealed that human resources analytics was not explored extensively; its usage was noted to be improving organizational performance and assisting in the alignment of people management initiatives and business goals. The important observation was that the benefits derived from the use of human resource analytics were commensurate with the size of the organization. The bigger organizations were observed to be better positioned to enjoy the benefits of human resource analytics. Boakye and Lamptey's (2020) research study conducted in Ghana revealed that people analytics assists the performance of organizations in the recruitment process positively influences productivity and employee retention. Most of the studied organizations lack HRA expertise, and some do not appreciate its use and application to business processes.

2.6 Zimbabwe-Specific Studies

A Zimbabwean research study (Mapuranga et al., 2024) in the tourism sector disclosed that a human resource information system, as an HR tool, improves various human resources functions in the tourism sector. It was further revealed that HRIS enables strategic business management, fast decision-making in the mining industry, and efficient rewards administration, which in turn motivates employees. The chances of employees leaving the organization unrecognized are avoidable because of the capabilities of strategic human resource management (Kajongwe, 2020). Mukuze (2023) research study on the adoption of human capital analytics in Zimbabwean state universities revealed that HRA was adopted but still in the infancy stage. There were funding challenges, resistance to change, and a lack of understanding of the subject matter by both HR professionals and support staff in the studied state universities in Zimbabwe. The above results were supported by Sixpence et al.'s (2023) research study, which revealed that a lack of mutual understanding, trust, and strategic fit between human resource strategies and business goals affected the adoption of human resource analytics. Mataruka et al.'s (2024) research study revealed that HRA leads to the achievement of long-term business goals, but it was also observed that managers need to have special digital skills for the adoption of HRA to be a success story.

Drawing from the human resource analytics perspectives presented above, it is apparent that there are knowledge and population gaps in the cited literature. The three perspectives (international, regional, and local studies) support the fact that studies were conducted in the mining, tourism, and small to medium enterprises, but there is a scarcity of the same research in tertiary education institutions, especially in Zimbabwe. There is an observation about the infancy stage of human resource analytics (Mukuze, 2023), but there is no properly suggested framework for the adoption of human resource analytics in Zimbabwean tertiary institutions. The above-cited research (Boakye & Lamptey, 2020; Mapuranga et al., 2024) clearly indicates that human resource analytics improved business performance and various human resources functions. Higher education institutions in Zimbabwe face peculiar challenges (associated with resistance to change, lack of necessary funds for the HRA operations, and lack of qualified personnel) that remain under-researched. There is a notable mismatch between people management strategies and general business goals (Sixpence et al., 2023), and there is no corrective framework that seeks to correct the anomaly in state universities.

The above research cited non-academic institutions, neglecting tertiary education institutions (Tessema et al., 2025; Sungwa, 2025). Though local studies by Kajongwe (2020) and Mapuranga et al. (2024) were focused on mining and tourism sectors, they still lack a tailored HRA model for the Zimbabwean state universities population. Yes, indeed, there is an acknowledgment of resistance to change by leaders and cultural barriers (Mukuze, 2023), but there is a lack of specific studies on how management capabilities and culture affect HRA adoption in tertiary institutions in Zimbabwe.

Therefore, it is justified to conduct the current study in Zimbabwean state universities because HRA was observed to improve general business performance (Bahuguna et al., 2023), and Zimbabwean universities still lag in the adoption of strategic human resource analytics. No previous studies were conducted to interrogate the performance of staff with strategic human resource analytics. The obstacles (money, expertise, and resistance) are distinct and call for solutions tailored to the Zimbabwean situation. There is no workable framework in the identified literature for integrating HR Analytics in higher education. The current study employs a mixed-methods approach to investigate the role of strategic human resources analytics in enhancing organizational performance to fill in the knowledge and population gaps mentioned above. The mixed-method research approach can close the research gaps created by the use of one research strand like qualitative research alone or quantitative research alone. Qualitative research findings lack the rigor found in quantitative research since the results are drawn from participants who give various truths about the study problem, and there is a subjective interpretation of the results. On the other hand, quantitative research results are rigid and lack the human elements like participants' views, suggestions, values, and opinions on the research problem (Creswell & Creswell, 2018). Hence, the use of a mixed-methods research strategy in this study would provide a strong base of results which enjoy the strengths of the two qualitative and quantitative research strands used in one study (Fetters et al., 2013).

2.7 Hypothesis Development

Drawing from the theoretical framework and empirical literature, the following hypothesis is proposed:

Hypothesis 1 (H1) There is a positive relationship between the adoption of SHRA and employee productivity in Zimbabwean state universities.

Encouraging Information: Studies indicate that organizations leveraging HR analytics report higher productivity, better talent retention, and improved decision-making (Bassil, 2011; Kiran et al., 2024).

3 Methodology

3.1 Research Philosophy

Pragmatism was the chosen philosophy due to its ability to include a plurality of investigation methods (Kaushik & Walsh, 2019). The major assumption of the pragmatism philosophy is that the meaning of human actions and beliefs is found in their consequences, and the world is not static, but rather always changing. Hence, action is particularly important in pragmatism school of thought. The study of strategic human resource analytics and organizational performance calls for the participants' views, opinions, and suggestions on all the possible benefits and challenges encountered in the day-to-day running of the business. The second part of the methodology sought to understand the organizational performance metrics, and this was only possible through objective measurement techniques. Hence, the use of the pragmatism worldview.

3.2 Research Approach

The researcher used a mixed-methods approach for the study on Strategic HR Analytics in Zimbabwean State Universities. A mixed-methods approach combines qualitative and quantitative data collection and analysis within a single study to provide a comprehensive understanding of the research problem (Creswell & Creswell, 2018). Qualitative methods in the form of interviews would assist in exploring HR analytics adoption levels, barriers, and perceptions in depth. On the other hand, quantitative methods (surveys, performance metrics) measure the statistical relationship between SHRA and organizational performance. Triangulation (use of the two research strands) strengthens validity by cross-verifying findings (Fetters et al., 2013). The strategy used was that qualitative interviews were held with HR professionals on SHRA challenges, and quantitative data were obtained from surveys of university staff on productivity impacts. The mixed methodology has empirical support of similar studies in African universities (Tessema et al., 2025), which used mixed methods to assess HR analytics adoption.

3.3 Research Design

The research design for the study was an embedded research design. An embedded research design involves simultaneously collecting and analyzing qualitative and quantitative data, then merging results to compare findings (Creswell & Plano Clark, 2018). The use of the embedded research design in this study is justified as its use provides both exploratory (qualitative) and confirmatory (quantitative) insights on SHRA. The specific objectives of the study are qualitative in nature, except for objective three, which sought to measure the relationship between strategic human resource analytics and employee productivity. The chosen design enhances data completeness by capturing numerical trends and narrative explanations. The research design was applied in such a way that the researcher conducted surveys (quantitative) and interviews (qualitative) at the same time and then compared results to identify consistencies or contradictions (survey data versus interview themes on SHRA barriers). The design has empirical support from previous studies like Kiran et al. (2024), which used an embedded research design in HR analytics research in India.

Table 2 Quantitative sampling method demonstration

Stratum	Population	Margin of error	Confidence level	Sample size
Academic staff	250	5%	95%	152
HR professionals	30	5%	95%	28
Administrators	20	5%	95%	20
Total	**300**			**200**

Source Primary Data (2025)

3.4 Sampling Methods

Purposive sampling method guided by the saturation principle was chosen to select participants who were directly involved in human resource analytics in the sampled state universities and had the subject matter knowledge of the discipline (Saunders et al., 2023). Accordingly, 6 HR professionals, 6 academic staff, and 4 administrators were purposely selected to make a total of 16 participants for the qualitative data. The sampling method ensures representation of key stakeholder groups in a research study.

For the quantitative research strand of the present study, the researcher chose stratified random sampling. The sampling method involves dividing the population into homogeneous subgroups (strata) and randomly selecting participants from each (Taherdoost, 2016). The sampling method ensures proportional representation of different university departments and reduces sampling bias compared to pure quota sampling. In the present study, the researcher divided the research respondents into academic staff, HR professionals, and administrators and then randomly selected respondents from each stratum using the online Raosoft calculator. Therefore, a total of two hundred respondents were obtained, overall, the sample size for the study was two hundred and sixteen (216) respondents. Boakye and Lamptey (2020) applied stratified sampling in Ghanaian HR analytics research. The sampling method is presented in Table 2.

3.5 Data Collection Tools

The researcher used two data collection tools, namely a semi-structured interview guide and the closed-ended research tool. A semi-structured interview guide is a flexible interview format with predefined themes but allows follow-up questions (Brinkmann, 2014). The instrument is justified in this study because it captures rich, contextual insights on SHRA challenges, and it is also adaptable to unexpected responses (for example, unique Zimbabwean barriers). A closed-ended questionnaire is a survey with fixed-response options, for example, Likert scales for quantitative analysis (Dillman et al., 2014). The instrument is justified in this study because it can measure the SHRA adoption levels statistically and enables regression analysis (for example, linking HR metrics to performance). Some of the research questions sought to understand the extent to which strategic human resource tools were adopted in Zimbabwean state universities. Questions related to the perceived relationship between the level of strategic human resource analytics adoption and employee productivity were also asked to the respondents. Other questions were about the most important barriers inhibiting the adoption of strategic human resource analytics adoption in state universities. Typical questions for the semi-structured interview guide were designed to explore how workforce data was being gathered and utilized in human resource decision-making processes. Potential limitations of the data collection processes were asked.

3.6 Data Synthesis Methods

The researcher used the thematic data analysis method for the qualitative research strand, as advised by scholars (Braun & Clarke, 2006), for interview transcripts. Descriptive and inferential statistical techniques were used for the quantitative research results of the study. The Statistical Package for Social Sciences (SPSS version 27) was the software used to assist with the statistical analysis of the data. Pearson product-moment correlation coefficient technique, and percentages were the descriptive statistics used to analyze demographic results of the study.

3.7 Ensuring Trustworthiness

The researcher employed several measures to achieve rigor and credibility of the qualitative and quantitative research findings for this study on Strategic HR Analytics in Zimbabwean State Universities. Some of the strategies employed in this study were drawn from scholarly evidence (Lincoln & Guba, 1985). The researcher employed a credibility strategy, which means the process of ensuring that research findings accurately reflect research subjects' views. One of the techniques applied was prolonged engagement in the field to understand the research context, and this was done for a period of approximately two months, collecting data, after the research permission letter was granted by the university authorities. The researcher also achieved triangulation by collecting data from distinct categories of employees, which include academic staff members, administrators, and HR professionals, to confirm the findings obtained. Research findings were shared with the research participants to validate their verbatim responses, and they confirmed the results as the true reflection of their understanding of the research subject matter (Birt et al., 2016). The researcher collected primary data for the research from participants who best represented the research phenomenon of HRA, which was purposive sampling to achieve transferability of the research findings.

3.8 Validity of Quantitative Data (Surveys and HR Metrics)

Validity in research refers to the accuracy of measurement, for example, ensuring that the data collection tool items cover the whole subject matter of the research. In this case, the researcher made sure that all the survey items covered the essential aspects of Strategic Human Resource Management (SHRA), thus achieving content validity for the study. Construct validity was ensured by ensuring that all the survey items measured the theoretical constructs, for example, HR Analytics adoption.

3.9 Reliability (Consistency of Measurement)

Internal consistency of the survey items was evaluated using Cronbach's Alpha (α). This was done to evaluate if the measures of survey items consistently assessed the same construct. A high Cronbach's alpha value indicates that the items are highly correlated and measure the same underlying construct, providing confidence in the survey's ability to capture the intended concept. A Cronbach's alpha measure of 0.7 is considered acceptable (Field, 2018). The ten-item scale demonstrated good internal consistency ($\alpha = .85$).

3.10 Ethical Considerations

Ethics in research means the ability to know what is right and wrong in human conduct and avoiding exposing humans to psychological and bodily harm during the research process. Ethical practices in a research study also entail properly acknowledging sources for ideas received from other scholars and following the designated referencing style. The American Psychological Association (APA) referencing style was followed throughout the process of authoring this book chapter. The researcher did not force the research subjects to participate in the study; instead, they were all told of the research process, its aim, and the implications of participating in the study. The researcher obtained informed consent from the participants, and they agreed to participate, knowing all the consequences of participating in the data collection process. Participants were also told of their right to withdraw from the study, and none withdrew. The researcher obtained research permission from the authorities in charge of the three organizations under study before the process of data collection. Confidentiality of the participants was maintained in the process of conducting interviews with participants. Participants were instructed not to mention their names in the interview transcripts, and pseudo names were adopted for the participants during the report writing process. In interviews, surveys were also anonymized, and coding of the quantitative data was followed.

4 Results and Discussion

4.1 Response Rate

From a sample of 216 research subjects, only 178 participants responded to the data collection tool, and this was a response rate of 82.4% of the sample size (178/216). The response rate is considered a good response rate according to Fincham (2008). The results are representative of the selected sample.

The age demographic results in Table 3 reveal that most of the respondents fall within the age range 35–44 years, and this makes approximately 31% of the chosen sample size. The second largest age range was 55 years and above (30.2%), and this depicts a picture of an older workforce in the studies of Zimbabwean state universities. The age range 25–34 years contributed approximately 10% of the sample size, and this age range means the young workforce is underrepresented in this study sample. It would mean that there is high turnover of this age group in state universities, or there could be early-career recruitment of the young people in this age category. The dominance of the older workforce in the state universities would mean possible resistance to digital workforce analytics adoption (Marler & Boudreau, 2017). On the positive side, it practically implies that the state universities may enjoy experience-driven human resource management policies. There is a signal of brain drain of younger employees and poor talent management in the state universities (Chinyamurindi, 2016).

Table 4 presents the chosen demographic elements of the respondents in terms of job category and educational level. From the primary data, academic staff have a master's degree and a doctorate as their highest educational qualifications (111 Master's, 9 Doctorates). This implies that state universities in Zimbabwe require strict academic qualifications for their varied faculties (Zishiri, 2025). On the other hand, it implies that academic staff would also support the adoption of data-driven decisions from their experience of research. Key staff in charge of human resource practices had both a diploma, bachelor's and master's degree qualifications (50% have a Master's (12/24), 25% Diploma (6/24), and 25% Bachelor's (6/24). However, they lacked the doctoral degree qualification. Practically, it means human resource

Table 3 Age range of respondents

		Frequency	Percent
Valid	25–34 years	16	9.9
	35–44 years	50	30.9
	45–54 years	47	29.0
	55 years and above	49	30.2
	Total	162	100.0

Source Primary Data (2025)

Table 4 Job category of staff * educational level of respondents

Count						
		Educational level of respondents				Total
		Diploma	Bachelor's degree	Master's degree	Doctorate degree	
Job category of staff	Academic staff	0	0	111	9	120
	HR Professionals	6	6	12	0	24
	Administrators	1	14	3	0	18
Total		7	20	126	9	162

Source Primary Data (2025)

practice roles demand practical experience over doctoral qualifications (Strohmeier & Piazza, 2015).

University staff employed as administrators were fairly educated as they had a diploma, a first degree and a master's degree. Table 4 reveals that 77.8% have a Bachelor's (14/18) or Diploma (1/18), and only 16.7% (3/18) hold a Master's. Scholars revealed that administrative duties can be executed well with general degrees even without higher degrees (Armstrong & Taylor, 2020).

4.2 *Hypothesis Testing*

Hypothesis 1 (H0): There is no relationship between the adoption of SHRA and employee productivity in Zimbabwean state universities.

Hypothesis 1 (H1): There is a positive relationship between the adoption of SHRA and employee productivity in Zimbabwean state universities.

The output in Table 5 shows the relationship between the two variables (strategic human resource analytics and its influence on employee productivity) under consideration in this study. The main interpretation is that there is a weak positive relationship between strategic human resource analytics and employee productivity $(r) = 0.031$. The correlation coefficient value was close to zero, which means there is almost no linear relationship between the two variables under consideration. Furthermore, there is no statistically $(0.349 > 0.05)$ significant relationship between the two variables. The analysis was based on a sample of 162 responses, and this is a reasonably large sample size. Hence, the primary data failed to reject the null hypothesis, which suggested that there is no correlation between the variables under study. There is no fundamental indication that strategic human resource analytics adoption determines worker productivity in the studied universities based on the obtained data.

The data were tested against the key assumptions of the Pearson product-moment correlation coefficient. The researcher ensured that all the variables were measured

Table 5 Correlations

		SHRA adoption influence employee productivity	Strategic Human Resource Analytics (SHRA) currently used
SHRA adoption influence employee productivity	Pearson Correlation	1	.031
	Sig. (1-tailed)		.349
	N	162	162
Strategic Human Resource Analytics (SHRA) currently used	Pearson Correlation	.031	1
	Sig. (1-tailed)	.349	
	N	162	162

Source Primary Data (2025)

on an interval scale of measurement. This was done by using Likert scale responses; the relationship between the variables was linear as observed through scatter plots. Table 6 shows the tests of normality.

Pearson product-moment correlation coefficient technique was tested for normality, and the Kolmogorov–Smirnov and Shapiro–Wilk tests were used in this study. Both tests were used to compare the normal distribution and significance level. SHRA adoption influence on productivity has Statistic = 0.170, $p = 0.000$ for Kolmogorov–Smirnov (K–S), and Statistic = 0.906, $p = 0.000$ for Shapiro–Wilk (S–W). The extent of SHRA currently used has Statistic = 0.438, $p = 0.000$ for Kolmogorov–Smirnov (K-S) and Statistic = 0.550, $p = 0.000$ for Shapiro–Wilk (S–W). Null Hypothesis (H_0): The data are normally distributed. Alternative Hypothesis (H_1): The data are not normally distributed.

The decision rule adopted was:

If p-value (Sig.) < 0.05, Reject H_0 (Data are not normal).

If p-value (Sig.) ≥ 0.05, Fail to reject H_0 (Data may be normal).

From the information above, the results are that both variables have $p = 0.000$ (< 0.05) in both tests; this shows compelling evidence against normality, and the data is significantly non-normal. Therefore, Pearson's correlation coefficient value above may be invalid. The bootstrapping procedure was conducted since bootstrapping

Table 6 Tests of normality

	Kolmogorov–Smirnov[a]			Shapiro–Wilk		
	Statistic	df	Sig.	Statistic	Df	Sig.
SHRA adoption influence employee productivity	.170	162	.000	.906	162	.000
Strategic Human Resource Analytics (SHRA) currently used	.438	162	.000	.550	162	.000

Source Primary Data (2025)
[a]Lilliefors Significance Correction

confidence intervals are not affected by the distribution of the scores; they are robust and provide accurate results.

As presented in Table 7, bootstrapping was performed with 1,000 resamples using simple random sampling, to assess the stability of the correlation coefficient, given non-normality. Key results were that bias = −0.001 indicating minimal systematic deviation from the original Pearson correlation (0.031), confirming that the initial estimate was not skewed. Standard Error (SE) = 0.040. A Confidence Interval (CI) containing zero aligns with the null hypothesis (H_0: $\rho = 0$), meaning no significant correlation exists (Field, 2018). The narrow range near zero further supports the lack of practical significance (Cohen, 1988). The negligible bias (−0.001) suggests that the original r was unbiased, while the SE (0.040) indicates moderate sampling variability (Mondy et al., 1993). The BCa 95% CI [−0.032, 0.099] and $p = 0.349$ reinforce that SHRA adoption does not meaningfully influence productivity in this context. This aligns with prior HR analytics literature suggesting that technology adoption alone (strategic human resource analytics (SHRA) may not enhance productivity without complementary factors like training, culture, or leadership (Marler & Boudreau, 2017).

Qualitative data were analyzed using the deductive thematic approach as revealed by Braun and Clark (2006). The analysis strategy followed the top-down approach

Table 7 Correlations—Bootstrapped

				SHRA adoption influence employee productivity	Strategic Human Resource Analytics (SHRA) currently used
SHRA adoption influence employee productivity	Pearson Correlation			1	.031
	Sig. (1-tailed)				.349
	N			162	162
	Bootstrap[a]	Bias		0	−.001
		Standard Error		0	.040
		BCa 95% Confidence Interval	Lower		−.032
			Upper		.099
Strategic Human Resource Analytics (SHRA) currently used	Pearson Correlation			.031	1
	Sig. (1-tailed)			.349	
	N			162	162
	Bootstrap[b]	Bias		−.001	0
		Standard Error		.040	0
		BCa 95% Confidence Interval	Lower	−.032	
			Upper	.099	

Source Primary Data (2025)
[a]Unless otherwise noted, bootstrap results are based on one thousand bootstrap samples

guided by the research objectives of this study and the need to maintain the exact words said by the research participants during the data collection process. Therefore, five main themes were drawn from the primary data gathered.

Theme 1: Level of Strategic Human Resource Analytics Employed in State Universities

Probing interview questions were asked to three categories of participants, and they gave their different views as presented below.

2 HR participants agreed that:

> *"At this university, we routinely collect workforce information that we use to run the payroll every month and supply similar headcount information to the ministry for the process of salary requests. There are no further analysis steps taken for the analysis of the gathered data besides the face value interpretation of the data."*

2 HR participants shared the view that:

> *"The language of analytics that you are talking about is foreign to our managers and section leaders; they regard the process of detailed analysis of people issues as the responsibility of the cost accounting section and not necessary to HRM personnel. Even myself, I just learnt about the word during my undergraduate studies, but in practice, I have not done any task that clearly shows how the aspect of strategic human resource management is done."*

2 HR participants averred that:

> *"The availability of advanced software would make the job of strategic human resource analytics very easy, for example analyzing retention rates and improving recruitment strategies. However, the organization here does not have good analytics software."*

Academic Staff (6 participants)

1 Academic staff participant:

> *"From my experience, decisions are made arbitrarily and not data-driven; my teaching duties are not at all related to HR Analytics."*

1 Academic staff participant:

> *"If training needs are identified through human resources data, then academic staff could improve research output and take the name of the organization to higher levels."*

4 Academic staff participants agreed as follows:

> *"The norm in this organization is that individual staff member promotions are done based on seniority and not associated with any performance metrics. The process of promotions based on seniority discourages highly performing academic staff members. The only metric used to award research incentives is the number of publications done by an individual member of staff. Though there are some research output incentives, some members do not receive them on time or do not receive them at all, and no reason is given by the people in charge of the process."*

Administrators (4 Participants)
2 ADM participants:

"Lack of adequate funds prevents us from acquiring workforce analytics suitable software. "

2 ADM Participants:

"The organization should have properly trained HR data analysts; at the moment, we have various reports being made but they are lacking proper analysis to give a proper and advanced picture of organizational standing."

The above results reflect the general picture that strategic human resource analytics is underutilized, as most of the human resources activities are just basic and routine work involving payroll processing and workforce attendance management. Therefore, it implies that the universities' understudy cannot make evidence-based decisions due to the absence of analytics. The findings corroborate Mukuze's (2023) research study on the adoption of human capital analytics in Zimbabwean state universities, which revealed that HRA was adopted but still in the infancy stage. There were funding challenges, resistance to change and a lack of understanding of the subject matter by both HR professionals and support staff in the studied state universities in Zimbabwe. The above results were supported by Sixpence et al. (2023) research study which revealed that a lack of mutual understanding, trust, and strategic fit between human resource strategies and business goals affected the adoption of human resource analytics. Mataruka et al.'s (2024) research study revealed that HRA leads to the achievement of long-term business goals, but it was also observed that managers need to have special digital skills for the adoption of HRA to be a success story.

Theme 2: Essential Human Resources Metrics Utilized to Enhance Organizational Performance

HR Professionals (6 Participants)
3 HR participants:

"In certain cases, we do headcount analysis and analyze payroll data, but not to a greater detail. When there is high staff turnover of lecturers, we can do verification but not to a greater depth and there are no ways to prevent that challenge."

1 HR participant:

"Yeah, I know that measuring employee engagement and even training efforts can help improve organizational performance, but here we consistently monitor attendance rate such that we do not pay ghost workers."

2 HR participants agreed that:

"Recruitment costs analysis for academic staff is done because the university pays for their transport when they are hired from a distant geographical area. However, there is no proper analysis to identify if the hiring process is slow and how that would affect service provision to key stakeholders. Recruitment process can sometimes take up to three months."

Academic Staff (6 participants)
2 Participants agreed that:

"In this organization, worker promotions are determined by the length of service not actual performance, and if there was a good criterion for measuring research or teaching quality, some staff members would be motivated to produce more quality stakeholders."

2 Participants averred that:

"They send us to training workshops, yes but I do not remember any attempt to measure what we were taught at the workshops and their application in day-to-day duty execution. So, in short, there is no data available on how training workshops improve employee skills."

2 Participants agreed that:

"Academic staff performance evaluation by students is done repeatedly, but the information collected is simply filed away and not used immediately. I strongly believe that if that information is used immediately, it will assist service delivery in the organization since areas of improvement are noted, and action is taken quickly to close performance gaps."

Administrators (4 Participants)
2 Participants revealed that:

"Budget allocations per faculty and departments is done, the only challenge is a good analysis of how the budget allocated would translate to good teaching services or research output."

2 Participants agreed that:

"Basic HR reports are done here but there are no advanced analytics like predictive analytics to be able to predict staffing needs in the future. Hence, at times we struggle to replace senior academics when they leave the organization."

The above results indicate that basic human resources metrics like attendance rate and turnover of staff are tracked in the state universities but there is a lack of strategic analysis of the same. Therefore, critical people management metrics like engagement and rate of return on training are out of the picture and this limits performance transformation analysis in the organizations.

From the literature review, primary data findings are supported by Tessema et al. (2025) research results in Ethiopia, which revealed that human resources analytics was not explored extensively; its usage was noted to be improving organizational performance and assisting in the alignment of people management initiatives and business goals. The important observation was that the benefits derived from the use of human resource analytics were commensurate with the size of the organization. The bigger organizations were observed to be better positioned to enjoy the benefits of human resource analytics.

Theme 3: Influence of Strategic Human Resources on Employee Productivity
Research participants gave different views on their understanding of how strategic human resources influence productivity.

HR Professionals (6 Participants)
1 HR Participant:

"Even with the introduction of a new performance appraisal system, it would not work well if the human resource policies were not linked to productivity of the organization. It will be very difficult to gauge if an appraisal system is working well."

2 Participants:

"Academic staff members always complain that promotions are not based on clear data and policy framework. It is like guesswork when it comes to HR decisions, and this challenge lowers the morale of the staff."

3 HR Participants:

"Workload distribution should always be visited because some lecturers in the faculty of social sciences and business management have heavy workloads, resulting in burnout, while Departments like Science and Technology have very few students and less workload for the lecturers."

Academic Staff (6 Participants)
3 AS Participants:

Some lecturers must have four modules to teach while others have just two or even one. There is no use of data to allocate tasks to academic staff, especially from the human resources section. It is possible to increase productivity of the organization if assignment of duties is based on authentic business data.

1 AS Participant:

"Sometimes I give up on working hard because our efforts at times go unrecognized, as the authorities in this organization do not track research output and do not observe high performers."

2 AS Participants:

"There should be a proper analysis of why employees leave the organization such that fixing the problems would be easy. You never know what is causing them to leave the organization, it could be the open issue of low salaries or leadership challenges in various departments."

Administrators (4 Participants)
3 ADM Participants:

"Service delivery is bound to suffer severely when there is no staffing in line with students' enrollment because you will notice that some faculties have more students than others."

1 ADM Participant:

"It is difficult to justify budget requests by university departments if there is no use of human resource analytics."

The absence of fair workload distribution because of a lack of data-driven human resources practices pushes important employees out of the organizations and reduces their zeal to work hard. Practically, it means state universities should optimize strategic human resource analytics as they can boost employee morale, exercise good staffing, be able to reward high-performing employees and eventually boost their productivity. The obtained results concur with Boakye and Lamptey's (2020) research study conducted in Ghana which revealed that people analytics assist the performance of organizations in the recruitment process, positively influence productivity and employee retention.

Theme 4: Obstacles to Implementing Strategic Human Resource Analytics
HR Professionals (6 Participants)
2 HR Participants:

"Our biggest challenge is lack of adequate sources, especially funding and advanced software for data analytics in the organization. Top leaders seem not to be worried by this analytics thing than other projects."

3 HR Participants:

"There is no one human resource professional trained in data analytics, thus lack of skills is still a barrier to utilization of strategic human resource analytics even if the relevant software and tools were available."

I HR Participant:

"Our leaders in this organization prefer to use traditional ways of human resources practices and decision making. Any changes brought by technology are heavily resisted by top managers who are used to the old ways of working."

Academic Staff (6 Participants)
1 AS Participant:

"This organization is still stuck in old ways of thinking that human resource management is about payroll and recruitment only but sideline the important aspect of strategies."

4 AS Participants:

"Top management in this university does not absolutely support strategic human resource management utilization. As usual, no business initiative would be successful without top leadership support."

1 AS Participant:

"There is no uniform data collection process in different university departments and that challenge renders proper data analysis useless."

Administrators (4 Participants)
2 ADM Participants:

"The university does not have good Information Technology base and some of the human resources systems are outdated and cannot successfully handle data analytics."

2 ADM Participants:

"There is no definitive policy on the use of human resource analytics in the organization. I believe it can only work if there is a policy and monitoring mechanism to ensure its application just like accounting and purchasing functions in the organization."

Drawing from the above verbatim reports, three main challenges affecting strategic human resource analytics can be categorized as skills gap, resistance to change, and shortage of funds. The best way to overcome such challenges would be to formulate water-tight policies and guidelines, identify staff training and leadership support. Therefore, the results concord with Sungwa's (2025) research study in South African Small to Medium Enterprises (SMEs), which disclosed that organizations understudy lack the services of properly qualified human resources professionals and the use of technology, which are all the essential requirements for the adoption of human resource analytics.

Theme 5: Framework for Incorporating Strategic Human Resource Management into Higher Education Practices

HR Professionals (6 Participants).

2 HR Participants:

"There is a need to move from basic human resource metrics like rate of employee turnover to predictive analytics, thus, a phased approach to strategic human resource analytics would be a good strategy to fully adopt human resource analytics."

2 HR Participants:

"Just like mandatory institutional audits, there should be a government policy which should force state universities to adopt human resource analytics."

2 HR Participants:

"It is important to partner with some technology organizations to get access to low-priced analytics software or tools."

Academic Staff (6 Participants)

2 AS Participants:

"The head of university human resources management and his team should work with different faculties to identify important performance indicators in form of teaching evaluations and research impact."

2 AS Participants:

"Human resource analytics should be done by a central department chosen to do the standardized data collection across various university departments."

2 AS Participants:

"It is important to embark on the necessary training of human resources section staff."

Administrators (4 Participants)

2 ADM Participants:

"A directing framework should include essential aspects like funding for software and personnel training."

2 ADM Participants:

> *"Clear key performance indicators should be set in the organization. This includes measuring success, reducing staff turnover by a certain percentage over a given period of time."*

The key picture shown by the results above is that strategic human resource analytics adoption should be done following the phased approach that considers policy formulation and support, relevant analytics training, and staggered implementation. Eventually, this tailor-made SHRA model would assist state universities to bridge the gap between current human resource practices and organizational performance. The results are in tandem with Nowicka et al. (2024) and Diefenhardt et al. (2025) research studies which disclosed that judicious use of data analytics in people management improves personnel matters and assists in the achievement of strategic goals of the organization. Al-Shammari et al.'s (2025) study results disclosed that poor data quality hinders the full utilization of strategic human resource analytics in organizations across the globe.

The primary data partially supported the theoretical framework for this study. The key theories used in this study assume that strategic human resource analytics should enhance workforce performance, but there are real-world challenges like culture, funding and skills gaps which prevent the positive influence of SHRA in Zimbabwean state universities. The numeric primary data (weak correlation) is in tandem with non-numeric data which disclosed that strategic human resource analytics is not systematically being applied in state universities. This explains why strategic human resource analytics impact on productivity is negligible. The non-numerical data provided contextual reasons why strategic human resource analytics remain untapped due to funding, cultural barriers, and leadership challenges.

5 Implication of the Study

The study results reported in this book chapter have theoretical, practical, policy, and methodological implications. The results extend the strategic human resource analytics literature to Zimbabwean state universities, where it was observed that the subject matter is under-explored, and the adoption barriers were different from those in other corporate business organizations. The results refine the human capital theory, which advocates for the training of human capital. In this case, results revealed that mere investment in human capital analytics is not sufficient, as organizations should also look at structural and cultural barriers affecting the adoption of SHRA. Practical implications for the study are that university top leadership needs to give priority to the training program for human resource analytics to improve their digital skills. There is a need for a phased approach to human resource analytics execution. Ministry of Higher Education officials in Zimbabwe should make SHRA adoption mandatory in state universities by enacting guiding frameworks and policies and allocating sufficient funds for software acquisition. Private–public partnerships with technology companies which provide low-cost technology tools and technical support are also important. Human resource professionals need to carry out benchmarking exercises against global best practices and have a mindset shift from old ways of doing business and integrate analytics into decision-making. While numeric data disclosed what is happening and non-numeric data explained why it was happening, future research should combine more data collection tools and varied HR metrics for stronger causal logical thinking.

6 Limitations and Areas for Further Research

There is a limitation on the generalization of the study results to other state universities or smaller tertiary institutions since only three state universities were studied. The qualitative perception of the aspect of employee productivity may not reflect the objective measurement of performance metrics in organizations. Cross-sectional research design could not establish causality in the failures of human resource analytics adoption; thus, the need to track the strategic human resource analytics influence on organizational performance over a long time.

Future studies can be conducted on the strategic human resource analytics adoption with universities in countries like South Africa versus Zimbabwe to identify the best regional practices. Such studies would explore affordable human resource analytics tools for resource-constrained universities and explore how leadership can reduce the issue of resistance to change in the adoption of strategic human resource analytics. It is of paramount importance that future studies look at how mandated and government-backed strategic human resource analytics adoption works in terms of adoption rates and productivity improvement.

7 Conclusion

Finally, the research study reveals that Zimbabwean state institutions of higher learning lag in strategic human resource analytics due to a plethora of challenges like financial restrictions, skill gaps, and cultural opposition. Nevertheless, participants acknowledge the possibility of improving productivity, equity, and important decision-making. A government-backed model, combined with training and low-cost technology, could drive SHRA integration. The study results provide a contribution to global human resource analytics literature in tertiary institutions. Future research should explore cost-effective analytics tools and change-management strategies for higher education.

References

Ackoff, R. L. (1971). Towards a system of systems concepts. *Management Science, 17*(11), 661–671.

Al-Shammari, M., Al Bin Ali, F., AlRashidi, M., & Albuainain, M. (2025). Big data and predictive analytics for strategic human resource management: A systematic literature review. *International Journal of Computing and Digital Systems, 17*(1), 1–9. https://doi.org/10.12785/ijcds/1571015706

Armstrong, M., & Taylor, S. (2020). *Armstrong's handbook of human resource management practice* (15th ed.). Kogan Page.

Bahuguna, P. C., Srivastava, R., & Tiwari, S. (2023). "Human resources analytics: where do we go from here?" *Benchmarking: An International Journal.*

Bassil, L. (2011). Raging debates in HR analytics. *People and Strategy, 34*(2).

Becker, G. S. (1964). *Human Capital: A theoretical and empirical analysis.* University of Chicago Press.

Bersin, J. (2023). *AI Unlocks New Power for Employees: Are HR Leaders Ready?* MIT Sloan Management Review.

Birt, L., Scott, S., Cavers, D., Campbell, C., & Walter, F. (2016). Member Checking. *Qualitative Health Research, 26*(13), 1802–1811. https://doi.org/10.1177/1049732316654870

Boakye, A., & Ayerki Lamptey, Y. (2020). The rise of HR analytics: Exploring its implications from a developing country perspective. *Journal of Human Resource Management, 8*(3). https://doi.org/10.11648/j.jhrm.20200803.19

Braun, V., & Clarke, V. (2006). Using thematic analysis in psychology. *Qualitative Research in Psychology, 3*(2), 77–101.

Brinkmann, S. (2014). Unstructured and semi-structured interviewing. In P. Leavy (Ed.), *The Oxford Handbook of Qualitative Research* (pp. 276–299). Oxford University Press. https://doi.org/10.1093/oxfordhb/9780199811755.013.030

Chinyamurindi, W. T. (2016). A narrative investigation into the meaning and experience of career success: Perspectives from women participants. *SA Journal of Human Resource Management, 15.* https://doi.org/10.4102/sajhrm.v14i1.659

Creswell, J. W. (2014). *Research design: Qualitative, quantitative and mixed methods approaches* (4th ed.). Sage.

Creswell, J. W., & Creswell, J. D. (2018). *Research design: Qualitative, quantitative, and mixed methods approaches.* Sage.

Creswell, J. W., & Plano Clark, V. L. (2018). *Designing and conducting mixed methods research.* (3rd ed.). SAGE.

Diefenhardt, F., Rapp, M. L., Bader, V., & Mayrhofer, W. (2025). 'In God We Trust. All Others Must Bring Data': Unpacking the influence of human resource analytics on the strategic recognition of human resource management. *Human Resource Management Journal, 35*(3), 597–612. https://doi.org/10.1111/1748-8583.12583

Dillman, D. A., Smyth, J. D., & Christian, L. M. (2014). *Internet, phone, mail, and mixed-mode surveys: The tailored design method* (4th ed.). John Wiley & Sons, Inc.

Di Prima, C., Cepel, M., Kotaskova, A., & Ferraris, A. (2024). Help me help you: How HR analytics forecasts foster organizational creativity. *Technological Forecasting and Social Change, 206,* Article 123540.

Drucker, P. F. (1954). *The practice of management.* Harper & Row.

Fetters, M. D., Curry, L. A., & Creswell, J. W. (2013). Achieving integration in mixed methods designs—principles and practices. *Health Services Research, 48*(6pt2), 2134–2156. https://doi.org/10.1111/1475-6773.12117

Field, A. (2018). *Discovering Statistics Using IBM SPSS Statistics* (5th ed.). Sage Edge.

Gallup. (2023). *State of the Global Workplace: 2023 Report.* Gallup, Inc.

Kajongwe, C. (2020). Transforming human resources in the digital age and performance of employees in the mining sector in zimbabwe and implications for development. *International Journal of African Business Studies (IJABS), 1*(2), 17–28.

Kaushik, V., & Walsh, C. A. (2019). Pragmatism as a research paradigm and its implications for social work research. *Social Sciences, 8*(9), 255. https://doi.org/10.3390/socsci8090255

Kiran, P. R., Chaubey, A., & Shastri, R. K. (2024). Role of HR analytics and attrition on organisational performance: A literature review leveraging the SCM-TBFO framework. *Benchmarking: An International Journal, 31*(9), 3102–3129. https://doi.org/10.1108/BIJ-06-2023-0412

Lengnick-Hall, M. L., et al. (2018). *Human resource management in the 21st century: Strategic HR analytics*. SAGE Publications.

Lincoln, Y. S., & Guba, E. G. (1985). *Naturalistic inquiry*. SAGE.

Mapuranga, R., Muzvondiwa, E., & Shateyi, S. (2024). Automation and human resources management in the tourism and hospitality industry of Zimbabwe. In *Tourism and Hospitality for Sustainable Development* (pp. 139–154). Springer Nature Switzerland. https://doi.org/10.1007/978-3-031-63077-4_8

Mataruka, L., Zishiri, C., Muzurura, J., Mkumbuzi, W. P., & Garatsa, C. (2024). Enhancing performance of zimbabwean service-based firms through digital transformation and sustainable competitive advantage: A structural equation analysis. *African Journal of Commercial Studies, 5*(1), 7–20. https://doi.org/10.59413/ajocs/v5.i1.2

Marongedza, L., Hlungwani, P. M., & Hove, P. (2023). Institutional constraints affecting secondary school student performance: A case study of rural communities in Zimbabwe. *Cogent Education, 10*(1). https://doi.org/10.1080/2331186X.2022.2163552

McCombes, S. (2023, January 2). *How to write a literature review|guide, examples, & templates*. https://www.scribbr.com/methodology/literature-review/.

McCombes, S. (2025, August). *What is a research design | types, guide & examples*. Scribbr.

McCombes, S. (2025, January 22). *How to write a literature review|guide, examples, & templates*. Scribbr. Retrieved June 3, 2025, from https://www.scribbr.com/dissertation/literature-review/

Mondy, R. W., Noe, R. M., & Premeaux, S. R. (1993). *Human resource management* (Fifth). Allyn and Bacon.

Muchowe, M. R., & Mubango, H. (2024). Human resources analytics in SMEs in Harare CBD: Challenges and opportunities. *Asian Macroeconomic Review, 4*(1).

Mukuze, K. (2023). *Developing a predictive model for human capital analytics adoption in Zimbabwean State Universities*. Chinhoyi University of Technology.

Ngwendere, J. S., & Mlitwa, N. (2025). Exploring the determinants of advanced big data analytics adoption in Zimbabwe's telecom sector: A TOE framework analysis. *Journal of Information Systems and Informatics, 7*(1), 356–377. https://doi.org/10.51519/journalisi.v7i1.1008

Nowicka, J., Pauliuchuk, Y., Ciekanowski, Z., Falda, B., & Sikora, K. (2024). The use of data analytics in human resource management. *European Research Studies Journal, XXVII* (Issue 2), 203–215. https://doi.org/10.35808/ersj/3380

Peter, L. J., & Hull, R. (1969). *The Peter Principle: Why things always go wrong*. William Morrow.

Ravitch, S. M., & Riggan, M. (2016). *Reason & Rigor: How conceptual frameworks guide research*. SAGE.

Saunders, M. N. K., Lewis, P., & Thornhill, A. (2023). *Research methods for business students* (9th ed.). Pearson.

Schindler, P. S. (2023). *Business research methods* (14th ed.). McGraw Hill.

Sixpence, S., Tapera, J., Ukpere, W., & Adekanmbi, F. (2023). Resetting the human resource strategy for organizational performance in Zimbabwe urban councils. Annals of Spiru Haret University. *Economic Series, 23*(2). https://doi.org/10.26458/23215

Strohmeier, S., & Piazza, F. (2015). *Artificial Intelligence Techniques in Human Resource Management—A Conceptual Exploration* (pp. 149–172). https://doi.org/10.1007/978-3-319-17906-3_7

Sungwa, J. (2025). Strategic human resources management practices are key to small, medium and micro enterprises effectiveness in South Africa. *SA Journal of Human Resource Management, 23*(0). https://doi.org/10.4102/sajhrm.v23i0.2763

Taherdoost, H. (2016). Sampling methods in research methodology: How to choose a sampling technique for research. *International Journal of Academic Research in Management, 5*(2), 18–27.

Tessema, S. A., Yang, S., & Chen, C. (2025). The effect of human resource analytics on organizational performance: Insights from Ethiopia. *Systems, 13*(2), 134. https://doi.org/10.3390/systems13020134

Webster, J., & Watson, R. T. (2002). Analyzing the past to prepare for the future: Writing a literature review. *MIS Quarterly.*

Zishiri, C. (2025). Quality dilemmas in higher education in zimbabwe: Qualitative perspectives. *International Journal of Research and Innovation in Social Science, IX* (IIIS), 2453–2461. https://doi.org/10.47772/IJRISS.2025.903SEDU0191

Kebiat Mukuze is an academic staff member at Midlands State University, Gweru main campus. Since 2020, he has focused 90% of his work on lecturing and research. He teaches Quantitative Research Methods and Labor Productivity at the undergraduate level, and Research Methodology and Statistics at the master's level. Kebiat Mukuze also serves as a Guest Editor for the Dyke Journal and as an external examiner at the University of Venda in the Republic of South Africa. He holds a DPhil in Human Resource Management, Chinhoyi University of Technology; a Master of Science in Human Resource Management, Great Zimbabwe University; a Post Graduate Diploma in Tertiary Education, Midlands State University; a BCom (Honors) in Human Resource Management, Zimbabwe Open University; and a Higher Diploma in Human Resource Management, Institute of People Management of Zimbabwe.

The Influence of AI Threats on Employee Behavior and Turnover Intentions

Shikha Saloni and Neema Gupta

Abstract This chapter explores the relationship between the Threat of Artificial Intelligence (TAI) and employee turnover intention, emphasizing the roles of role ambiguity and knowledge hiding behavior. This study employs a quantitative approach using PLS-SEM to analyze how perceived Artificial Intelligence (AI) threats influence turnover intentions. The findings indicated that the Threat of AI (TAI) had a significant positive impact on Turnover Intention ($\beta = 0.289, p < 0.001$), Role Ambiguity ($\beta = 0.368$, $p < 0.001$), and Knowledge Hiding Behavior ($\beta = 0.670, p < 0.001$). Additionally, Role Ambiguity ($\beta = 0.119, p < 0.01$) and Knowledge Hiding Behavior ($\beta = 0.259, p < 0.01$) significantly increased turnover intention, while the mediating effect of role ambiguity (H4) was not supported. The implications of these findings highlight the need for organizations to foster transparent communication and supportive environments to reduce role ambiguity and prevent knowledge hiding. By actively engaging employees in discussions about AI, clearly defining job roles, and promoting a culture of knowledge sharing, organizations can mitigate turnover intentions. This study contributes and offers insights for future research to examine additional factors influencing employee responses to technological changes.

Keywords Threat of AI · Turnover Intention · Role Ambiguity · Knowledge Hiding Behavior · PLS-SEM

1 Introduction

Turnover Intention (TI) refers to the level at which employees think about quitting their organization voluntarily and is a major concern for employers around the world (Guzeller & Celiker, 2019). High turnover rates are usually accompanied by

S. Saloni (✉)
School of Management, ITM University, Gwalior, India
e-mail: shikhasaloni.som@itmuniversity.ac.in

N. Gupta
Graphic Era Deemed to be University, Dehradun, India
e-mail: neema.e11856@cumail.in

M. Dzingirai and S. O. Atiku (eds.), *Strategic Human Resource Analytics and Intelligent Data-Driven Systems*, https://doi.org/10.1007/978-981-92-2623-8_3

higher costs of recruiting, training, and the loss of organizational memory. Besides, turnover affects key organizational routines, leading to reduced morale and productivity (Verma & Kesari, 2019). As competition intensifies, talent management has emerged as a critical human resource management agenda hitting "tipping point" across the globe (Hom et al., 2017). Studies in organizational behavior also reveal many more aspects that cause turnover intention: job dissatisfaction, poor leadership, and stress. Nonetheless, the modern issue that has attracted considerable discussion is the effects of Artificial Intelligence (AI) on employment trends. Automation, a core component of AI, has increased efficiency but simultaneously displaced many low-skill roles (Mhlanga, 2021). This duality of efficiency versus employment vulnerability drives much of the turnover anxiety explored in this study. The related concept of Threat of AI (TAI), defined as the perceived fear and vulnerability that employees may experience as a result of being managed or displaced by AI systems, has been identified as connected with turnover intention more and more (Jiang & Probst, 2021). Workers who believe that AI is a threat to their jobs will feel more insecure in their jobs, and hence they will look for other jobs. This has been experienced especially where companies are introducing technologies that automate tasks that involve data processing and repetitive tasks, which employees cannot easily understand their next line of duty.

While the direct association of Threat of AI and turnover intention is indisputable, this relation is frequently influenced by additional factors. Two such notable factors in this regard are Knowledge Hiding Behavior (KHB) and Role Ambiguity (RA). These factors proactively explain how employees perceive the threat of AI in their workplace and give a deeper understanding of the antecedent process of turnover intention (Başer et al., 2024). Clearly, when employees see AI as a threat to their livelihood, then they will develop possessiveness over knowledge while maintaining important skills to ensure they are not replaced (Beckett, 2021). Such behavior in turn negatively affects trust within implementing teams, cooperation, and augmentation of feelings of isolation, leading to turnover intention (Serenko & Bontis, 2016). Knowledge hiding, the purposeful act of keeping knowledge that should help others or the organization, is a behavior known to be influenced by both environmental conditions and individual concerns (Connelly et al., 2012). In the context of AI, employees might conceal information since they think that submitting that information might lead to the elimination of their positions quickly. Besides knowledge hiding, there is another significant mediator between AI threat and turnover intention, and it is role ambiguity. Role ambiguity is a situation where the workers are in doubt about their roles, their tasks, or the standards that will be used to appraise them. The introduction of AI technologies can be a challenge to existing job descriptions and employees often find themselves wondering if and how their work tasks will be affected or how they will need to be reconfigured (Turel et al., 2018). When AI is introduced to an organization and is not yet fully understood, it can lead to stress and anxiety among the employees due to unknown ways of operating that come along with the change in their working environment.

Despite a recent surge in scholarly interest in the effects of AI on employee turnover, several key questions have not been answered to a satisfactory extent. First, despite the negative effects of AI on turnover intention and job insecurity have been documented (Choi et al., 2023; Jiang & Probst, 2021), little is known regarding the various psychological mechanisms, particularly knowledge hiding, that explain this relationship. Further research is required to establish comprehension of how the threat of AI results in preventive measures such as knowledge hiding and how this influences an employee's turnover decision (Serenko, 2023). Second, there is a lack of research regarding the impacts of role ambiguity in the case of AI adoption. While the integration of AI technologies in teams may lead to overlapping of the roles and thus, there is more uncertainty about responsibilities, there is still very limited empirical research focusing on the effects of this ambiguity on people's turnover intentions (Turel et al., 2018). It was also emphasized that comprehension of how AI adoption results in role creation or deterioration is important for reducing its adverse influence on work environments, especially in innovative areas (Kim et al., 2020).

Thus, through this chapter, we aim to address the following research objectives:

Objective 1: To analyze the effect of threat of AI incorporation and automation on employees' turnover intention.
Objective 2: To determine the effect that Role ambiguity has on the relation of threat of AI and Turnover Intention of employees.
Objective 3: To establish whether the knowledge hiding behavior of the employee has any impact on the relationship between AI and turnover intention among the employees.

Thus, by advancing our research aims, we will contribute to the existing body of literature investigating the effects of technology-anticipated changes on employees. The structure of this chapter is as follows: First, the literature on the main ideas in our research model is discussed; then, the data and procedures are described. Research results of the empirical study are next discussed, followed by the discussion, conclusion, and suggestions for further research.

2 Literature Review

2.1 Threat of Artificial Intelligence and Turnover Intention

Poberznik et al. (2021) provided an example of how technological advancements can reduce the risk of employee turnover, particularly in the event of a pandemic. Kareem et al. (2020) findings support a similar claim, showing that utilizing a technology-based Customer Relationship Management (CRM) system can remarkably lower employees' TI. In order to enhance employee retention, Eickemeyer et al. (2021) suggested i4.0, a software tool that can detect possible drawbacks of digitization on workers. Although some research indicates that integrating AI could lower turnover intentions, there is conflicting data in the literature. Ayyagari et al. (2021) have reported that there are studies that suggest artificial intelligence (AI) may contribute to a rise in quit intentions as a result of the additional stress that comes with using Information and Communication technologies (ICTs). This is referred to as techno-stress. Work pressure escalation, perception of increased workload, work lethargy, frustration, and lack of motivation are some of the issues that act as consequences of technological stressors. They can also result in decreased job satisfaction, increased TI, and a lack of commitment to organization (Vuori et al., 2019). The feeling of dissatisfaction from a job has been proven to be an antecedent of TI, which is a strong precursor to actual turnover (Johnsrud & Rosser, 2020). Thus, the following hypothesis was proposed:

H1: The threat of AI positively impacts employees' turnover intention.

2.2 Role Ambiguity as a Mediator between Threat of AI and Turnover Intention

Role ambiguity is likely more common in organizations where AI integration is inconsistent or not well-communicated. AI can thus demotivate employees because they lack adequate knowledge and clarity concerning where and how they are useful, are unclear on which roles or tasks are pertinent, and how they will be appraised. In similar situations, the workers receive a perception of losing organizational mastery over their careers, thus becoming dissatisfied, and are more likely to seek other jobs (Kim et al., 2020). AI implementation always changes tasks, which leads to organizational roles and responsibilities and the measures of performance not being well defined (Turel et al., 2018). The extent of the impact of automating the different tasks performed by the employees is not clear to them, and this creates confusion about their futures within the organization. Such ambiguity creates stress, anxiety, and dissatisfaction, which in turn leads to turnover intention (Kim et al., 2020). When one lacks a clear understanding of how an organization will use AI in their workplace, this category of employees will feel that they have little control over their careers in the organization and, therefore, will seek new roles in other organizations with clear career advancement paths. However, if employees are not told how their jobs are going to change due to AI-based automation, then this may intensify role ambiguity and, therefore, increase job insecurity, as identified in a study by Choi et al. (2023). This often results from the fast-growing application of AI solutions and ambiguous expectations; employees are forced to look for jobs that clearly outline their duties and where human input exceeds the effectiveness of an artificial solution.

H2: Threat of AI significantly impacts role ambiguity.

H3: Role ambiguity positively influences employees' TI.

H4: Role ambiguity acts as a mediator between the association of threat of AI and TI.

2.3 Knowledge Hiding Behavior as a Mediator between TAI and TI

Knowledge hiding refers to the intentional concealment of information or expertise when others request it, often to maintain one's job security or perceived value (Connelly et al., 2012). Within AI-driven workplaces, this behavior becomes a defensive resource-protection strategy. Scholarly research on knowledge hiding indicates that the main factors influencing workers' knowledge hiding behavior are knowledge characteristics, organizational-level variables, and individual-level variables (He et al., 2021). Previous studies have indicated that employees' hesitation to divulge their knowledge to others can occasionally be driven by a defensive mindset and a wish to keep custody of the ownership of their knowledge (Huo et al., 2016). As intelligent machinery becomes more prevalent in the workplace and employees begin to perceive it as a threat to their jobs and careers, they will take defensive measures. These may include acting carelessly, concealing information, and operating as though they are unaware of the problems they are being asked to solve (Bari et al., 2020; Feng & Wang, 2019). However, even though they feel a hostile environment and encounter these conflicts, the collaborators will exercise caution and selectiveness when sharing information. Particularly, they will try to postpone sharing critical information as long as possible to establish themselves as essential and irreplaceable members of the team (Serenko & Bontis, 2016). Alternatively, they will fabricate a story to support their decision to withhold information (Kumar Jha et al., 2018; Ma et al., 2020). A company culture that encourages withholding information can be harmful to both the employees and the company. A few problems linked to KH include decreased creativity (Chen et al., 2020), interpersonal and organizational deviance (Singh, 2019), decreased mutual trust (Cerne et al., 2017; Yao et al., 2020), and turnover intention (Anand et al., 2020; Sheidaee et al., 2022).

H5: Threat of AI significantly impacts employee's KHB.

H6: KHB positively influences employees' TI.

H7: KHB acts as a mediator between the association of the threat of AI and TI.

2.4 Theoretical Framework

2.4.1 Stress-Strain Outcome Model

The stress-strain outcome model can be adopted to understand the impact the threat of AI has on the turnover intention. In this context, the existence of AI becomes a stressor since it leads to the uncertainty of the job and future roles among employees (Brougham & Haar, 2020). This uncertainty produces psychological pressure, which results in such behaviors as knowledge hiding and the emergence of role ambiguity

(Pereira & Mohiya, 2021). By failing to add clarity toward employees' expectations and role descriptions, they also enhance pressure levels (Xu & Xue, 2023). Knowledge hiding transpires due to the perceived risk that proffering proficiency might cause the initiation of redundancy concerning the employees. The state of role ambiguity is an indication of the confusion that employees experience while trying to adapt to AI-mediated changes within their work (Singh, 2019; Yao et al., 2020). When these strains build up, they can lead to such outcomes as intention to turnover. Seymour and Hartman (2012) postulated that people who are highly stressed at work are likely to disengage at work and feel that they are not connected with their workplace and, thus, may search for other jobs where their roles and responsibilities seem more secure to them (Li et al., 2019; Turel et al., 2018). This type of disconnect may sometimes lead to an increase in turnover intention as employees tend to look for workplace environments that provide them with greater stability and clarity in their roles (Kim et al., 2020).

2.4.2 Conservation of Resources (COR) Theory

Conservation of resources (COR) theory, which was postulated by Hobfoll (1989), postulates that individuals do all within their abilities to gain, develop, and protect resources that are valuable to them, and which may encompass the person, conditions, and energy. When it comes to the threat of AI and the turnover intent, the COR framework explains how employees manage threats and stressors at the workplace. When employees think that their jobs are threatened by AI, they are likely to experience a high level of stress, which drains their resources (Halbesleben & Buckley, 2004). These assets may include resources that can be touched, including employment and income, but also things that cannot be touched, like self-esteem, job satisfaction, and the ability to identify with the job (Gorgievski & Hobfoll, 2008).

When gaining knowledge regarding the risks connected with the use of AI in the organization, employees may act in ways that protect their expertise, which translates into hiding knowledge within the organization in an attempt to ensure their perceived value in the organization (Bari et al., 2020; Connelly et al., 2012). This behavior, although meant to secure their resources, can create a culture of distrust and reduced cooperation among the workers at the workplace (Cerne et al., 2017). However, it is worth mentioning that stress rises with role ambiguity because the lack of clear roles in an AI-driven environment can lead to further resource loss (Kahn et al., 1964; Lee et al., 2021a, 2021b). Such a strain may lead to a higher turnover intention since employees desire to avoid exposure to an environment that is destructive to their resources and health (Hom et al., 2017; Jiang & Probst, 2021). According to COR theory (Hobfoll, 1989), individuals strive to conserve valuable resources such as skills, job stability, and social capital. When employees perceive AI as a threat, they may engage in knowledge hiding as a coping mechanism to protect these resources. This behavioral response demonstrates COR's applicability beyond stress management, extending into defensive knowledge behaviors in technologically disruptive contexts.

3 Research Methodology

3.1 Research Design

This study employed a quantitative, cross-sectional research design to examine the relationships between Threat of Artificial Intelligence (TAI), Role Ambiguity (RA), Knowledge Hiding Behavior (KHB), and Turnover Intention (TI) among employees. Data were collected using a structured questionnaire measured on a five-point Likert scale ranging from 1 (Strongly Disagree) to 5 (Strongly Agree). To test the proposed hypotheses and mediation relationships, partial least squares structural equation modeling (PLS-SEM) was used. This method was considered appropriate due to its suitability for analyzing complex models involving multiple constructs and mediation effects.

3.2 Target Population and Sample Size

The target population consisted of employees working in different organizations across various industries in North India who may experience the impact of AI adoption in the workplace. Respondents were working professionals aged between 25 and 55 years. This study was done using a structured questionnaire with a Likert scale of strongly disagree (SD) = 1, disagree (D) = 2, neutral (N) = 3, agree (A) = 4, and strongly agree (SA) = 5 from June 2024 to August 2024. To determine the bare minimum value of the subjects required for this research, the G*Power software was used, as proposed by Faul et al. (2009). The minimum sample size was 159 when statistical power would be at 0.80 and the alpha level at 0.05.

3.3 Data Collection

In this study, an adapted survey was administered to be filled out by various participants working for different organizations. All the responses were collected using the snowball sampling technique as recommended by Neuman (2006). Because this study involves a digital questionnaire conducted through E-mails and Whatsapp from colleagues, associates and friends, non-random Snowball sampling proves suitable because it is impossible to know the entire dispersed population of employed people (Gunawan & Huarng, 2015). Snowball technique was used to reach out to the respondents in the future, in this case, family members, friends and any employee in any organization was given the option to fill the questionnaires. Therefore, in total, about 490 people were contacted and were found to be aged between 25 and 55 years of age. With an overall response rate of 90%, there was an overall total of 456 responses

Table 3.1 Demographic data

Profile	Characteristics	Frequency	Percentage (%)
Gender	Male	170	42.6
	Female	229	57.3
Income	0–3 lakhs	96	24.0
	3–6 lakhs	83	20.8
	6–9 lakhs	161	40.3
	Above 9 lakhs	59	14.7
Age	25–35	106	26.5
	36–45	189	47.3
	46–55	104	26.0
Marital status	Married	287	71.9
	Unmarried	112	28.1

Source Author's Own

from the respondents. Out of 441 useful responses after the validation of the data and deleting multivariate outliers of 15 responses, the data evaluation was done.

3.4 Demographic Data

The demographic data of the respondents are presented in Table 3.1.

3.5 Measures

Data from participants were gathered using a modified questionnaire that was based on the research model illustrated (see Fig. 3.1). There are two main components to the questionnaire's finalized form. Age, gender, annual income, and marital status are just a few of the demographic factors that were included in the first section. The four constructs, namely Threat of AI, RA, KH, and TI, are measured in the second section. TAI, RA, KH, and TI were evaluated using Brougham and Haar (2018), Nübling et al. (2006), Connelly et al. (2011), and Menezes et al., respectively.

3.6 Common Method Bias

Selecting the appropriate question format while carrying out a survey is another factor that should be considered, since it is often subject to Common Method Bias (CMB). This mainly happens when information is collected from one source (Avolio et al., 1991). Full collinearity was used to test the intra-correlation of the Variance Inflation Factor (VIF) to investigate the actual common method variance (Kock, 2015). Out of

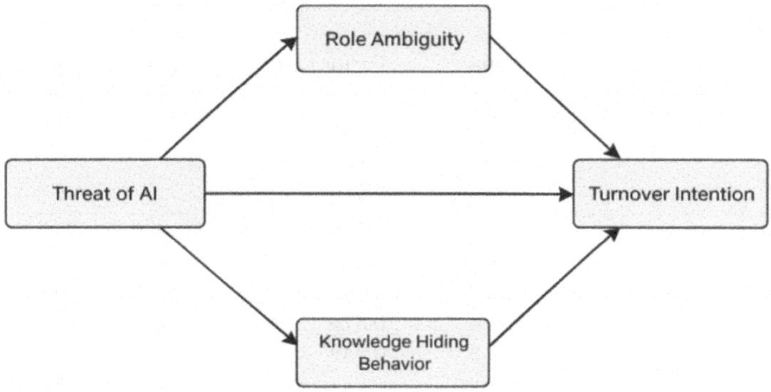

Fig. 3.1 Conceptual model. *Source* Author's Own

all the constructs, the maximum value of the VIF inner was 1.676, which was lower than the limit of 3.3. Consequently, CMB could not be flagged as an issue within the proposal for the study.

Standardized root mean square residual (SRMR) is used to assess how well a model fits data, and it should be less than 0.08 (Henseler et al., 2014). In the current model, the SRMR value is 0.068, which is less than the critical threshold of 0.08. The predicted correlation matrix of the research model accurately replicated the empirical correlation matrix, which met the acceptable critical SRMR value.

3.7 Measurement Model

The measurement model was evaluated using internal consistency reliabilities, composite reliability, and Henseler's rhoA for reflective constructs within the tolerance range of 0.70–0.95 (Hair et al., 2019, 2022) (Table 3.2). Average variance extracted was used to determine convergent validity; reflective constructs exceeded the 0.50 threshold. Except for three, almost all reflective constructs exceeded the 0.70 threshold, indicating indicator reliability (Hair & Alamer, 2022). Despite this, the previous three indicators were retained because their values exceeded 0.60. (Hair et al., 2022), and constructs had met the AVE threshold of 0.50 (Hair et al., 2019).

Heterotrait-Monotrait (HTMT) ratio of correlations was used to test the discriminant validity (Henseler et al., 2015) (Table 3.3). HTMT was below the tolerance limit of 0.85 (Gold et al., 2001). Thus, it is concluded that the current measurement model's discriminant validity is not impacted. The results of discriminant validity are presented in Table 3.3.

Table 3.2 Construct validity

Construct	Items	Factor Loading	Rho_A:rho_c	AVE
Threat of artificial intelligence	TAI 1	0.843	0.911: 0.936	0.784
	TAI 2	0.892		
	TAI 3	0.894		
	TAI 4	0.910		
Role ambiguity	RA 1	0.905	0.949: 0.958	0.820
	RA 2	0.860		
	RA 3	0.942		
	RA 4	0.879		
	RA 5	0.939		
Turnover intentions	TI 1	0.821	0.908: 0.932	0.775
	TI 2	0.891		
	TI 3	0.900		
	TI 4	0.907		
Knowledge hiding	KHB 1	0.699	0.934: 0.936	0.548
	KHB 2	0.714		
	KHB 3	0.722		
	KHB 4	0.723		
	KHB 5	0.697		
	KHB 6	0.722		
	KHB 7	0.806		
	KHB 8	0.799		
	KHB 9	0.802		
	KHB 10	0.789		
	KHB 11	0.680		
	KHB 12	0.718		

Source: Authors' Calculation

Table 3.3 Discriminant validity (HTMT ratio)

	KHB	RA	TAI
KHB			
RA	0.575		
TAI	0.701	0.395	
TI	0.555	0.389	0.546

Source Authors' Calculation

4 Results

4.1 Structural Model

The evaluation of the structural model was carried out in agreement with the recommendations made by Hair et al. (2019, 2022). The VIF inner values, which were discovered to be less than the critical value of 3.33 (Hair et al., 2019), were used to diagnose the collinearity issues. The structural model was evaluated based on the

Table 3.4 Structural model assessment

Hypothesis	Predictor variable	Path relationship	Path coefficient (β)	Confidence interval	VIF	f^2	Results
H1	TAI	TAI-TI	0.289***	[0.125; 0.431]	1.597	0.370	Yes
H2	TAI	TAI-RA	0.368***	[0.280; 0.458]	1.000	0.409	Yes
H3	RA	RA- TI	0.119**	[0.028; 0.213]	1.308	0.219	Yes
H5	TAI	TAI-KHB	0.670***	[0.605; 0.733]	1.000	0.813	Yes
H6	KHB	KHB—TI	0.259**	[0.104; 0.431]	1.796	0.352	Yes

Source Authors' Calculation
Note (*** = $p > 0.001$; ** = $p > 0.01$)

following metrics: path coefficients (β), R^2, F^2, and confidence interval for examining the causal relationships between the investigated latent variables (Hair et al., 2019, 2022). The variance in the model is explained by the R^2 value. With the provided R^2 for latent variables, the model exhibits good explanatory power, with R^2 values of TI = 0.322, KH = 0.448, and RA = 0.135. The value can be interpreted as 13.5% variance in Role Ambiguity, 44.8% variance in knowledge hiding and 32.2% variance in turnover intentions is determined by artificial intelligence (independent variable). Hypothesis testing using the bootstrapping method was followed by the previous step. Moreover, 10,000 subsamples of bootstrapping were used to evaluate the structural model results (Hair et al., 2022). Table 3.4 shows the findings of the structural model. H1, H2, and H3 are supported as a result of the findings that TAI has a significant and positive effect on TI ($\beta = 0.322$), KHB ($\beta = 0.448$), RA ($\beta = 0.135$), RA has a significant effect on TI ($\beta = 0.119$), and KHB has a significant and positive effect on TI ($\beta = 0.259$). The results are presented in Fig. 3.2.

4.2 *Mediation Analysis*

Hypotheses 4 and 7 state that RA and KHB act as conciliators in the association between workplace rudeness and intentions to leave, respectively. Table 3.4 reveals the mediation role of them. The Role Ambiguity and knowledge hiding-mediated relationship between Threat of Artificial Intelligence and intention to leave has a VAF value of 0.132 and 0.376, respectively, which explains no mediation and partial mediation, respectively. Table 3.5 shows the results from the mediation analysis.

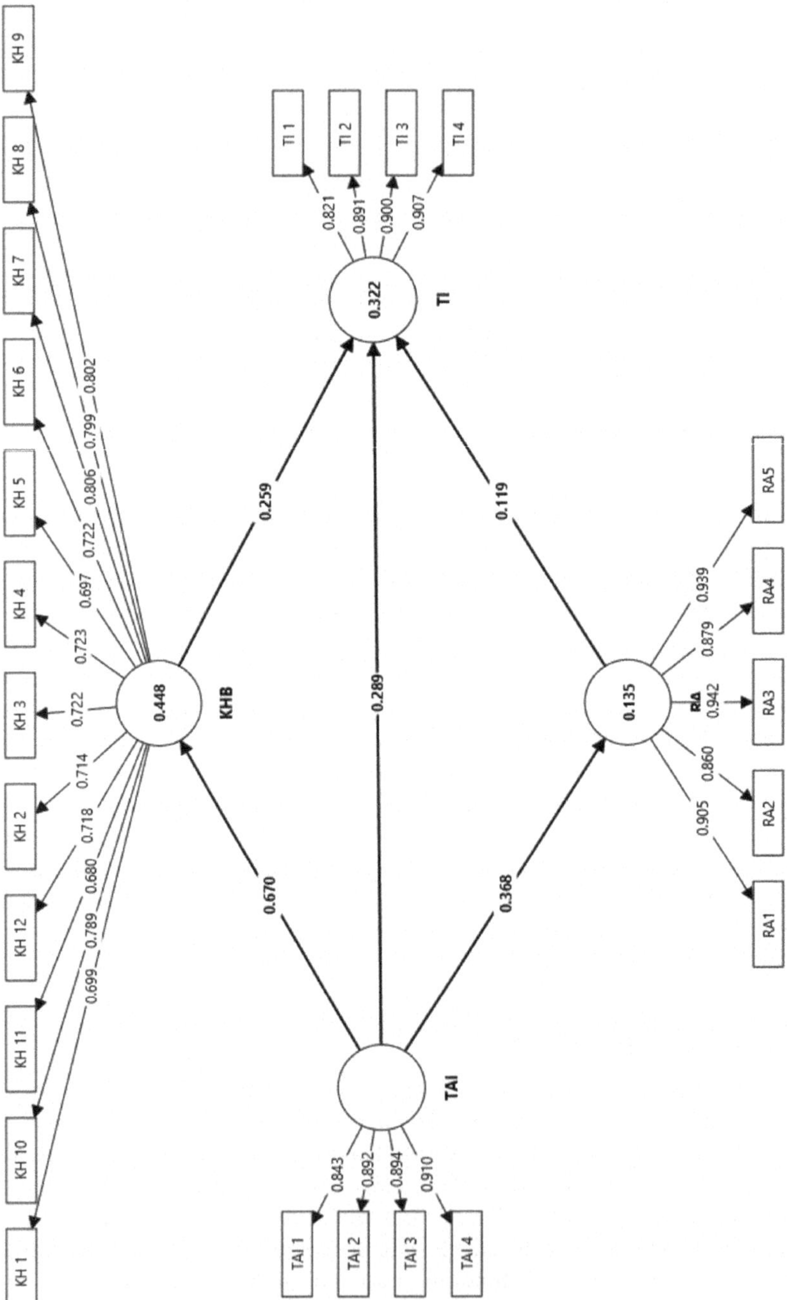

Fig. 3.2 Structural model

Table 3.5 Mediation analysis

Hypothesis	Path relation			VAF	Decision
H4	AI- RA-TI	DE IDE TE	0.289 0.044 0.333	0.132	NME (VAF < 0.20)
H7	AI- KHB-TI	DE IDE TE	0.289 0.174 0.462	0.376	PME (0.20 < VAF < 0.80)

Source Authors' Calculation

5 Discussion

The purpose of this research was to establish the role of AI threat on TI, RA, and KHB in various organizations. By employing a quantitative methodology and PLS-SEM in order to examine the stated hypotheses, several substantiated links appeared.

5.1 *Threat of AI and Turnover Intention*

The research results reveal that the threat of AI has a positive relationship with employee turnover intention (Hypothesis 1). This result supports other studies that have shown that, due to uncertainties arising from the adoption of new technology like AI, employees become less secure in their jobs, hence develop intentions of leaving their employers (Li et al., 2019; Moreira et al., 2020). With the advancement of AI in the work environment, employees fear the loss of their jobs or that their jobs will be automated, and therefore, the likelihood of exploring other career options or employers with strong assurances that their jobs are safe increases. It is aligned with the notion that automation can be detrimental to employees' sense of security and increases turnover intentions (Segovia-Perez et al., 2023). However, unlike earlier studies that primarily framed AI as a structural disruptor, the present research demonstrates that employees interpret AI threats through psychological mechanisms, such as perceived loss of control and diminished self-efficacy (Kim & Lee, 2025). Thus, this study extends existing literature by showing that subjective threat appraisal, not merely objective job changes, drives turnover intention.

5.2 *Threat of AI and Role Ambiguity*

The positive and strong correlation between the threat of AI and role ambiguity (Work Role Uncertainty) provided support for H2 and extended the literature in HRM, indicating the disruptive effects of AI on employees. Whenever new technology is integrated into working environments, it may result in high levels of uncertainty in work roles or even expectations for employed personnel and therefore leads to high levels of role ambiguity (Koo et al., 2021). This is more so where the AI is likely to alter the structural arrangements and skill demand in relation to jobs, creating uncertainty among the employees. This is in line with previous studies, which show that new technologies, such as AI, pose threats to employment structure and entail more uncertainty in the workplace (Xu & Xue, 2023). These findings expand on Xu and Xue (2023), revealing that ambiguity arises not only from unclear organizational design but also from emotional apprehension about losing professional identity. Consequently, AI-related change management should emphasize transparent task redistribution to preserve role clarity.

5.3 Role Ambiguity and Turnover Intention

For Hypothesis 3, this study revealed that role ambiguity is positively and significantly related to turnover intention (H3). This finding is consistent with the previous findings, which claim that when the employed workforce feels that their positions are uncertain, they are ready to search for another job (Gull et al., 2023; Arias-Pérez & Vélez-Jaramillo, 2021). Through much of this, it becomes clear that employees who are uncertain of their roles and responsibilities, or employees with blurred roles due to AI integration and implementation, will be more likely to quit and seek other organizations that offer them better-defined roles. Role ambiguity and turnover intention have been found to be related to one another for many years and are significant determinants of employee retention policies that most organizations consider, especially when experiencing technological transformations (Laily & Ernawati, 2020).

5.4 Mediation of Role Ambiguity

Although role ambiguity was hypothesized as a mediator between AI threat and turnover, the results showed no significant mediation effect (VAF < 0.20). Thus, H4 Stands Rejected. This diverges from Lee and Jeong (2017), who found partial mediation in contexts of technological uncertainty. The finding suggests that role ambiguity independently influences turnover but does not fully explain how the AI threat translates into withdrawal. Hence, other mediating factors, such as trust erosion or psychological safety, may better capture this linkage in AI-intensive contexts.

5.5 Threat of AI and Knowledge Hiding Behavior

The study also showed that the threat of AI has a positive impact on knowledge hiding behavior (H5). This is in line with previous research, which had found that when employees perceive threat from AI, they may resort to knowledge hiding to safeguard their employment (Sheidaee et al., 2022). A form of knowledge management that has previously been observed is knowledge hiding, whereby employees keep information or skills from others with the conception that such a course of action will protect them from being replaced by AI solutions (Jena & Swain, 2021). Such behavior is becoming more and more applicable to the organization that is going through the process of digital transformation, and where AI is considered a threat to one's work value.

5.6 *Knowledge Hiding Behavior and Turnover Intention*

The last Hypothesis 6 (H6), which stated that knowledge hiding has a direct and positive impact on turnover intention, was supported by the findings. As with studies before, those who practice knowledge hiding may experience feelings of organizational exile or organizational distrust, the latter of which leads to higher turnover intentions (Pereira & Mohiya, 2021). This may be a mix of insecurity or perceived threats from AI to their work. This would continue to push employees further away from their organizational obligations. When employees intentionally conceal information, they are also likely to distance themselves from the job, making them even more inclined to look for other job opportunities (Xiong et al., 2021). Unlike prior literature that treated knowledge hiding as an outcome, this study identifies it as a mechanism linking AI-induced fear to exit decisions, emphasizing its dual role as both a coping response and a risk amplifier.

5.7 *Mediation of Knowledge Hiding Behavior*

The partial mediation of knowledge hiding in the threat of AI and turnover intention (H7) offers valuable information on how threats associated with AI have an impact on employees' actions. Specifically, knowledge hiding is a partial mediator of this relationship with VAF of 0.20–0.80, indicating that, although directly AI boosts turnover intention, it also indirectly does so through knowledge hiding (Yao et al., 2020). These findings stress that defensive behaviors mediate the relationship between technological threats and employees' outcomes, suggesting that managers should tackle knowledge-sharing challenges in AI-intensive settings (Yao et al., 2020). While prior studies (Kim et al., 2020; Choi et al., 2023) emphasized AI's effect on job insecurity, our findings reveal that defensive behaviors like knowledge hiding partially mediate this link, suggesting psychological, not just structural, pathways of turnover.

6 Implications of the Study

6.1 Practical Implications

The findings of this study provide several practical implications for organizational leaders, policymakers, and human resource managers who are integrating artificial intelligence into their operations. The positive relationship between the threat of AI and employee turnover intention highlights the urgent need for transparent communication about the purpose, scope, and outcomes of AI implementation. Managers should clearly define new roles and performance expectations to minimize uncertainty and anxiety among employees. For example, in the banking and financial services sector, employees often fear that AI-driven analytics will replace their roles; however, open communication about how AI complements human decision-making can reduce perceived threats and foster acceptance. In addition, the partial mediation effect of knowledge hiding indicates that organizations must cultivate a culture of openness and trust. Leaders can achieve this by promoting collaborative platforms, encouraging team learning, and recognizing employees who share expertise rather than conceal it. For instance, in manufacturing or logistics industries, firms can implement "AI learning circles" or peer coaching programs that emphasize skill exchange and reinforce the message that AI adoption enhances, rather than replaces, human capability.

Moreover, the findings emphasize the critical role of AI reskilling and continuous learning in reducing turnover intention. HR departments should design targeted reskilling initiatives such as workshops on digital literacy, AI ethics, and data interpretation to ensure that employees remain confident in their evolving roles. In sectors like healthcare administration, where AI increasingly supports diagnostics and data management, structured training can reassure employees of their continued relevance and importance. By integrating such programs into career development pathways, organizations can transform fear of displacement into motivation for personal growth. Finally, organizations should adopt a holistic approach to employee well-being during AI transitions. This includes providing psychological support through counseling or mentoring, establishing clear communication channels for concerns, and addressing ethical issues related to automation. By investing in transparency, capability development, and emotional support, employers can convert the perceived threat of AI into an opportunity for innovation, engagement, and long-term retention. In doing so, organizations not only enhance workforce adaptability but also build sustainable human–AI collaboration that strengthens competitive advantage.

6.2 Theoretical Implications

In providing new perspectives on the threat of AI, this study adds to the existing knowledge regarding the relationship between AI and employee behavior in the following ways. The study of knowledge hiding as the partial mediator between AI threats and turnover intention contributes to filling a theoretical gap regarding defensive responses to technological change. Previous studies have centered on job insecurity and turnover intentions from more traditional viewpoints like leadership and culture (Brougham & Haar, 2020). In doing so, this work expands the literature on AI as a disruptive technology for analyzing current theories of organizational behaviors and responses to technological changes. Furthermore, the role of role ambiguity on the relationship between perceived AI threat and turnover intention implies that other factors may mediate these relationships. Future research could examine how specific factors, such as the support provided to the employee, which may be in the form of organizational support or training programs, or even the level of discretion allowed at the workplace to the employee, may moderate these threats relating to AI technologies (Jehanzeb, 2020). Altogether, this work may serve as a starting point for the exploration of the psychological and behavioral reactions of employees regarding the AI progress, which is important for scholars addressing organizational behavior and technology students.

7 Limitations and Future Scope

This study contributes to the knowledge pool by establishing the relationship between AI threats and turnover intention, role ambiguity, and knowledge hiding, but future research can extend this knowledge by paying attention to the following areas. First, it would be interesting for researchers to examine antecedents such as the context, organizational culture, leadership, and employee burnout as possible moderators of the relationship between AI and workplace behaviors. These factors may act as moderators or amplifiers of the undesirable effects related to AI integration, which gives a richer insight into how AI transitions can be approached and managed (Engström et al., 2024). However, future research should employ panel designs to capture changes in employees' perceptions of and actions towards threats of AI over time as organizations increasingly embrace AI technologies. It may assist in establishing the impact of AI adoption on turnover intention and the sharing of knowledge in the long run. The next research avenue that should be explored is the effects of training and development programs for AI on employees' job security and the propensity for knowledge hiding. Cross-cultural research would also help in finding out how the perception of AI varies from one nation to another or from one industry to another. Last, exploring the emotional and cognitive reactions, for example, stress or burnout, in relation to AI or the level of technological self-efficacy might be valuable to further understand the psychological factors leading to turnover intention and knowledge hiding. These future directions are essential to expand the theoretical and practical contributions of AI-related studies in various organizations and sharpen the understanding of AI implementation without adverse impacts on employees.

8 Conclusion

In contributing to the field of research related to the integration of AI in the workplace, this study explored how the threat of AI impacts both the turnover intention of employees and the level of role ambiguity and knowledge hiding behavior in the workplace. Quantitative method and PLS-SEM analysis of the results of the present study show that there are positive connections between the Threat of AI and turnover intention, as well as between the AI and knowledge hiding, supporting most of the research hypotheses. The findings indicate that AI creates a rational risk to employees, where they act defensively, which results in knowledge hiding and higher turnover intentions. Notably, the partial mediation of knowledge hiding between AI threats and turnover intention provides an understanding of how employees manage perceived threats to their employability. In contrast to the hypothesis, role ambiguity was not found to act as a Mediator between perceived threats from AI and turnover intention, which can indicate that other factors could be more influential in this process. The presented research results have a significant value for the development of both theoretically-based tools for effective implementation and practical solutions in organizations. From the theoretical point of view, the research moves previous literature on workplace incivility and knowledge hiding one step further by incorporating threats associated with AI into the topic. This reveals the importance of bringing in a finer appreciation of the way new technologies call for changes to the patterns of embodied interaction in the workplace. From a management perspective, the findings are useful for organizational leaders who help their firms to navigate decisions for implementing AI. Thus, increasing transparency, decreasing role ambiguity, and valuing knowledge sharing give a possible way for managers to eliminate the adverse influence of AI on the retention of employees. The results also stress the need for retraining and reskilling strategies in order to minimize threats of job loss, as well as assist employees in transitioning to automated advances. Overall, this study highlights that AI adoption, while inevitable, must be paired with transparent communication, clear role definition, and employee reskilling to preserve workforce trust and reduce turnover intentions.

References

Anand, A., Walsh, I., & Moffett, S. (2020). Knowledge hiding and turnover intention: A three-wave investigation of key employee attitudes. *Journal of Business Research, 111*, 281–292. https://doi.org/10.1016/j.jbusres.2019.12.024

Arias-Pérez, J., & Vélez-Jaramillo, J. (2022). Understanding knowledge hiding under technological turbulence caused by artificial intelligence and robotics. *Journal of Knowledge Management, 26*(6), 1476–1491.

Avolio, B. J., Yammarino, F. J., & Bass, B. M. (1991). Identifying common methods variance with data collected from a single source: An unresolved sticky issue. *Journal of Management, 17*(3), 571–587.

Ayyagari, R., Grover, V., & Purvis, R. (2021). Techno-stress: Technological antecedents and implications. *MIS Quarterly, 45*(1), 75–103. https://doi.org/10.25300/MISQ/2021/15479

Bari, M. W., Abrar, M., Shaheen, S., Bashir, M., & Fanchen, M. (2020). Knowledge hiding behaviors and team creativity: The contingent role of perceived mastery motivational climate. *Journal of Knowledge Management, 24*(9), 1–24. https://doi.org/10.1108/JKM-04-2020-0244

Başer, M. Y., Büyükbeşe, T., & Ivanov, S. (2024). The effect of STARA awareness on hotel employees' turnover intention and work engagement: The mediating role of perceived organisational support. *Journal of Hospitality and Tourism Insights*. https://doi.org/10.1108/jhti-12-2023-0925

Beckett, D. (2021). The death of human capital? Its failed promise and how to renew it in an age of disruption. *International Journal of Lifelong Education, 40*(3), 297–301. https://doi.org/10.1080/02601370.2021.1949178

Bhargava, A., Bester, M., & Bolton, L. (2020). Employees' perceptions of the implementation of robotics, artificial intelligence, and automation (RAIA) on job satisfaction, job security, and employability. *Journal of Technology in Behavioral Science, 6*(1), 106–113. https://doi.org/10.1007/s41347-020-00153-8

Brougham, D., & Haar, J. (2018). Smart technology, artificial intelligence, robotics, and algorithms (STARA): Employees' perceptions of our future workplace. *Journal of Management & Organization, 24*(2), 239–257. https://doi.org/10.1017/jmo.2016.55

Brougham, D., & Haar, J. (2020). Technological disruption and employment: The influence on job insecurity and turnover intentions: A multi-country study. *Technological Forecasting and Social Change, 161*, 120276. https://doi.org/10.1016/j.techfore.2020.120276

Cao, G., Duan, Y., Edwards, J. S., & Dwivedi, Y. K. (2021). Understanding managers' attitudes and behavioral intentions towards using artificial intelligence for organizational decision-making. *Technovation, 106*, 102312. https://doi.org/10.1016/j.technovation.2021.102312

Cerne, M., Nerstad, C. G. L., Dysvik, A., & Škerlavaj, M. (2017). What goes around comes around: Knowledge hiding, perceived motivational climate, and creativity. *Academy of Management Journal, 57*(1), 172–192. https://doi.org/10.5465/amj.2012.0122

Chen, J., Ma, L., Jin, J., & Gao, W. (2020). Knowledge hiding: A process model and theoretical investigation. *Knowledge Management Research & Practice, 18*(1), 87–98. https://doi.org/10.1080/14778238.2019.1604228

Choi, K., Park, J., Lee, J., & Kim, T. (2023). The impact of artificial intelligence on employment and the labor market: A focus on job insecurity and turnover intention. *Journal of Business Research, 144*, 51–61. https://doi.org/10.1016/j.jbusres.2022.10.006

Connelly, C. E., Zweig, D., Webster, J., & Trougakos, J. P. (2011). Knowledge hiding in organizations. *Journal of Organizational Behavior, 33*(1), 64–88.

Connelly, C. E., Zweig, D., Webster, J., & Trougakos, J. P. (2012). Knowledge hiding in organizations. *Journal of Organizational Behavior, 33*(1), 64–88. https://doi.org/10.1002/job.737

Eickemeyer, D., Klein, A., & Hommel, R. (2021). i4.0 tool for evaluating the effects of digitization on employee retention. *Journal of Business & Industrial Marketing, 36*(2), 223–238. https://doi.org/10.1108/JBIM-03-2020-0127

Engström, A., Pittino, D., Mohlin, A., Johansson, A., & Mirzaei, N. E. (2024). Artificial intelligence and work transformations: Integrating sensemaking and workplace learning perspectives. *Information Technology and People*. https://doi.org/10.1108/itp-01-2023-0048

Faul, F., Erdfelder, E., Lang, A.-G., & Buchner, A. (2009). GPower 3: A flexible statistical power analysis program for the social, behavioral, and biomedical sciences. *Behavior Research Methods, 39*(2), 175–191. https://doi.org/10.3758/bf03193146

Feng, J., & Wang, C. (2019). Exploring the dark side of workplace friendships: The moderating effects of envy and employee emotional intelligence on knowledge hiding and job performance. *Journal of Knowledge Management, 23*(6), 1330–1354. https://doi.org/10.1108/JKM-12-2018-0734

Gold, A. H., Malhotra, A., & Segars, A. H. (2001). Knowledge management: An organizational capabilities perspective. *Journal of Management Information Systems, 18*(1), 185–214. https://doi.org/10.1080/07421222.2001.11045669

Gorgievski, M. J., & Hobfoll, S. E. (2008). Work context as a resource: The role of work-related resources in the relationship between work and family stress. *Journal of Occupational Health Psychology, 13*(3), 265–278. https://doi.org/10.1037/1076-8998.13.3.265

Gull, A., Ashfaq, J., & Aslam, M. (2023). AI in the workplace: Uncovering its impact on employee well-being and the role of cognitive job insecurity. *International Journal of Business and Economic Affairs, 8*(4), 79–91.

Gunawan, J., & Huarng, K.-H. (2015). Viral effects of social network and media on consumers' purchase intention. *Journal of Business Research, 68*(11), 2237–2241. https://doi.org/10.1016/j.jbusres.2015.06.004

Guzeller, C. O., & Celiker, N. (2019). Examining the relationship between organizational commitment and turnover intention via a meta-analysis. *International Journal of Culture Tourism and Hospitality Research, 14*(1), 102–120. https://doi.org/10.1108/ijcthr-05-2019-0094

Hair, J., & Alamer, A. (2022). Partial Least Squares Structural Equation Modeling (PLS-SEM) in second language and education research: Guidelines using an applied example. *Research methods in applied linguistics, 1*(3), 100027.

Hair, J. F., Hult, G. T. M., Ringle, C. M., & Sarstedt, M. (2019). *A primer on partial least squares structural equation modeling (PLS-SEM)* (2nd ed.).

Hair, J. F., Hult, G. T. M., Ringle, C. M., Sarstedt, M., & Thiele, K. O. (2022). Mirror, mirror on the wall: A comparative evaluation of composite-based structural equation modeling methods. *Journal of the Academy of Marketing Science, 50*(6), 1375–1404. https://doi.org/10.1007/s11747-022-00846-0

Halbesleben, J. R. B., & Buckley, M. R. (2004). Burnout in organizational life. *Journal of Management, 30*(6), 859–879. https://doi.org/10.1016/j.jm.2004.06.004

He, P., Xu, Z., Wang, W., & Zhou, W. (2021). Knowledge hiding in organizations: Current research status and future research directions. *Journal of Organizational Behavior, 42*(2), 97–118. https://doi.org/10.1002/job.2475

Henseler, J., Dijkstra, T. K., Sarstedt, M., Ringle, C. M., Diamantopoulos, A., Straub, D. W., ... & Calantone, R. J. (2014). Common beliefs and reality about PLS: Comments on Rönkkö and Evermann. *Organizational research methods, 17*(2), 182–209.

Hom, P. W., Lee, T. W., Shaw, J. D., & Hausknecht, J. P. (2017). One hundred years of employee turnover theory and research. *Journal of Applied Psychology, 102*(3), 530–545. https://doi.org/10.1037/apl0000103

Hobfoll, S. E. (1989). Conservation of resources: A new attempt at conceptualizing stress. *American Psychologist, 44*(3), 513.

Huo, W., Cai, Z., Luo, J., Men, C., & Jia, R. (2016). Antecedents and intervention mechanisms: A multi-level study of knowledge hiding in teams. *Journal of Organizational Behavior, 37*(5), 1–23. https://doi.org/10.1002/job.2081

Jehanzeb, K. (2020). Does perceived organizational support and employee development influence organizational citizenship behavior? *European Journal of Training and Development, 44*(6/7), 637–657. https://doi.org/10.1108/ejtd-02-2020-0032

Jena, L. K., & Swain, D. (2021). How knowledge-hiding behavior among manufacturing professionals influences functional interdependence and turnover intention. *Frontiers in Psychology, 12*, 723938.

Jiang, L., & Probst, T. M. (2021). The threat of automation: Job insecurity and turnover intention in the face of AI and automation. *Computers in Human Behavior, 123*, 106877. https://doi.org/10.1016/j.chb.2021.106877

Johnsrud, L. K., & Rosser, V. J. (2020). Faculty members' morale and their intention to leave: A multilevel explanation. *Journal of Higher Education, 81*(2), 1–21. https://doi.org/10.1080/00221546.2020.1036459

Kahn, R. L., Wolfe, D. M., Quinn, R. P., Snoek, J. D., & Rosenthal, R. A. (1964). *Organizational stress: Studies in role conflict and ambiguity*. John Wiley & Sons.

Kareem, S. M., Saeed, K. O., & Haider, K. (2020). Effectiveness of technology-based CRM systems in reducing turnover intention: Evidence from IT industries. *Technological Forecasting and Social Change, 151*, 119808. https://doi.org/10.1016/j.techfore.2019.119808

Kim, B. J., & Lee, D. G. (2025). Self-efficacy in using artificial intelligence as a shield: Mitigating the detrimental effects of organizationally prescribed perfectionism on employee stress and anxiety. *Current Psychology, 44*(3), 1805–1831.

Kim, T., Tam, L., & Lee, J. (2020). AI adoption and role ambiguity: How to reduce its impact on employee turnover in innovative areas. *Journal of Organizational Change Management, 33*(5), 753–772. https://doi.org/10.1108/JOCM-02-2020-0035

Kivimäki, M., Nyberg, S. T., Batty, G. D., & Fransson, E. I. (2018). Job strain and employee turnover: The role of age. *Journal of Occupational Health Psychology, 23*(4), 469–478. https://doi.org/10.1037/ocp0000102

Kock, N. (2015). Common method bias in PLS-SEM: A full collinearity assessment approach. *International Journal of E-Collaboration (ijec), 11*(4), 1–10.

Koo, B., Curtis, C., & Ryan, B. (2021). Examining the impact of artificial intelligence on hotel employees through job insecurity perspectives. *International Journal of Hospitality Management, 95*, 102763.

Laily, N., & Ernawati, D. P. (2020). The effect of knowledge sharing and innovation behavior on the performance of batik entrepreneurs. *Jurnal Manajemen, 24*(2), 194–209.

Lee, S. H., & Jeong, D. Y. (2017). Job insecurity and turnover intention: Organizational commitment as mediator. *Social Behavior and Personality: An international Journal, 45*(4), 529–536.

Lee, J. Y., Choi, J., & Park, H. (2021a). Emotional intelligence, knowledge hiding, and creativity: The mediating effect of workplace relationships. *International Journal of Human Resource Management, 32*(9), 2043–2065. https://doi.org/10.1080/09585192.2019.1591291

Lee, Y., Chang, K., & Park, J. (2021b). Role ambiguity and turnover intention: The mediating effect of job stress. *International Journal of Environmental Research and Public Health, 18*(3), 1141. https://doi.org/10.3390/ijerph18031141

Li, W., Fang, Y., & Jia, F. (2019). Artificial intelligence, technological innovation, and employee turnover intention. *Computers in Human Behavior, 98*, 17–29. https://doi.org/10.1016/j.chb.2019.04.008

Liu, H., & Liu, Y. (2023). The mediating role of job insecurity in the relationship between artificial intelligence adoption and turnover intention. *Technology in Society, 72*, 102109. https://doi.org/10.1016/j.techsoc.2023.102109

Meyer, J. P., & Allen, N. J. (1997). Commitment in the workplace: Theory, research, and application. *Sage*. Publications.

Mhlanga, D. (2021). Artificial intelligence in the industry 4.0, and its impact on poverty, innovation, infrastructure development, and the sustainable development goals. *Lessons from Emerging Economies? Sustainability, 13*(11), 5788. https://doi.org/10.3390/su13115788

Michalos, A. C., & Tsaousis, I. (2016). Knowledge hiding and performance in the workplace: A new conceptualization and future research directions. *Journal of Knowledge Management, 20*(6), 1144–1164. https://doi.org/10.1108/JKM-04-2016-0178

Moreira, A., Cesário, F., Chambel, M. J., & Castanheira, F. (2020). Competences development and turnover intentions: the serial mediation effect of perceived internal employability and affective commitment. *European Journal of Management Studies, 25*(1), 5–22.

Nasir, M. H., & Kaur, J. (2020). The impact of emotional intelligence on knowledge sharing behavior: The moderating effect of trust in colleagues. *Journal of Knowledge Management, 24*(9), 2105–2125. https://doi.org/10.1108/JKM-11-2019-0620

Neuman, W. L. (2006). Social research methods: Qualitative and quantitative approaches (Vol. 30 No. 3). Pearson Education Limited. https://doi.org/10.2307/3211488

Nübling, M., Stößel, U., Hasselhorn, H. M., Michaelis, M., & Hofmann, F. (2006). Measuring psychological stress and strain at work-Evaluation of the COPSOQ Questionnaire in Germany. *GMS Psycho-Social Medicine, 3*, Doc05.

Park, J., Choi, K., & Kim, J. (2022). The role of job stress in turnover intentions: An analysis of the moderating effects of workplace relationships. *Journal of Managerial Psychology, 37*(4), 246–261. https://doi.org/10.1108/JMP-09-2021-0472

Pereira, V., & Mohiya, M. (2021). Knowledge hiding and job insecurity: Implications for employee turnover intentions. *Journal of Business Research, 135*, 622–632. https://doi.org/10.1016/j.jbusres.2021.07.032

Poberznik, A., Leino, M., Huhtasalo, J., Jyräkoski, T., Valo, P., Lehtinen, T., ... & Virkki, J. (2021). Mobile robots and rfid technology-based smart care environment for minimizing risks related to employee turnover during pandemics. *Sustainability, 13*(22), 12809.

Razi, M. A., & Sami, M. F. (2021). The mediating role of knowledge sharing in the relationship between ethical leadership and employee performance. *Journal of Business Ethics, 173*(2), 265–279. https://doi.org/10.1007/s10551-020-04554-4

Reitz, A. W., & Schmidt, G. (2022). Bridging the gap between artificial intelligence and human resources: A conceptual framework for AI implementation in HRM. *International Journal of Human Resource Management.* https://doi.org/10.1080/09585192.2022.2106934

Ruck, K., & Welch, M. (2012). Evaluating the effectiveness of internal communication: A case study approach. *Public Relations Review, 38*(5), 121–134. https://doi.org/10.1016/j.pubrev.2012.03.006

Salanova, M., & Schaufeli, W. B. (2008). A cross-national study of work engagement as a mediator between job resources and turnover intention. *Journal of Business Research, 61*(1), 124–132. https://doi.org/10.1016/j.jbusres.2007.05.008

Schaufeli, W. B. (2013). What is engagement? In S. G. Rogelberg (Ed.), *The Oxford handbook of workplace stress* (pp. 70–89). Oxford University Press.

Segovia-Perez, M., Jianu, B., & Tussyadiah, I. (2023, January). Assessing turnover intentions of algorithmically managed hospitality workers. In *ENTER22 e-Tourism conference* (pp. 349–354). Cham: Springer Nature Switzerland.

Serenko, A. (2023). The great resignation: the great knowledge exodus or the onset of the great knowledge revolution? *Journal of Knowledge Management, 27*(4), 1042–1055.

Serenko, A., & Bontis, N. (2016). Understanding counterproductive knowledge behavior: antecedents and consequences of intra-organizational knowledge hiding. *Journal of Knowledge Management, 20*(6), 1199–1224.

Seymour, D. E., & Hartman, J. L. (2012). Workplace stress: The role of workload, job control, and social support in predicting employee satisfaction. *Journal of Occupational Health Psychology, 17*(2), 213–225. https://doi.org/10.1037/a0027300

Sharma, R., & Singla, R. (2019). Knowledge hiding and creativity: The moderating role of workplace ostracism. *Journal of Knowledge Management, 23*(6), 1325–1335. https://doi.org/10.1108/JKM-06-2018-0382

Sheidaee, S., Philsoophian, M., & Akhavan, P. (2022). The effect of intra-organizational knowledge hiding on employee turnover intentions: the mediating role of organizational embeddedness: a case study of knowledge workers of IRIB. *Journal of Organizational Effectiveness: People and Performance, 9*(3), 422–448.

Singh, S. K. (2019). Knowledge hiding in organizations: Causes, consequences, and moderating factors. *Journal of Organizational Behavior, 40*(6), 783–800. https://doi.org/10.1002/job.2399

Spector, P. E. (1997). Job satisfaction: Application, assessment, cause, and consequences. *Sage.* Publications.

Spenner, P. M., & Freeman, R. B. (2016). Why people work: The role of employee engagement in the workplace. *Harvard Business Review, 94*(9), 66–72.

Sweeney, P. (2022). AI and the future of work: Exploring the impact of artificial intelligence on employee turnover. *Business Horizons, 65*(3), 367–376. https://doi.org/10.1016/j.bushor.2021.10.004

Vuori, J., Törnroos, K., Ruokolainen, M., & Wallin, M. (2019). Enhancing late-career management among aging employees–a randomized controlled trial. *Journal of Vocational Behavior, 115*, 103327.

Turel, O., Connelly, C. E., & Zhu, L. (2018). AI and turnover intention: How job roles, stress, and ambiguity contribute to employee decisions. *Journal of Occupational and Organizational Psychology, 91*(2), 365–387. https://doi.org/10.1111/joop.12206

Verma, B. K., & Kesari, B. (2019). Does the morale impact on employee turnover intention? An empirical investigation in the Indian steel industry. *Global Business Review, 21*(6), 1466–1488. https://doi.org/10.1177/0972150919856957

Wognum, A., & Kraaijenbrink, J. (2023). AI-induced job insecurity and turnover intention: The mediating role of job satisfaction. *European Journal of Work and Organizational Psychology, 32*(2), 284–298. https://doi.org/10.1080/1359432X.2022.2091768

Wu, C. H., & Bilgihan, A. (2018). The role of innovation and human capital in creating high-performance work systems. *International Journal of Human Resource Management, 29*(5), 1015–1037. https://doi.org/10.1080/09585192.2017.1280192

Xiong, R., Wei, P., Yang, J., & Cristofini, L. A. (2021). Impact of top executive turnover on firms' R&D investment: evidence from China. *Innovation, 23*(3), 400–424.

Xu, H., Hu, M., & Chen, S. (2022). How AI impacts employees' turnover intentions: The mediating roles of job stress and organizational commitment. *Frontiers in Psychology, 13*, 830260. https://doi.org/10.3389/fpsyg.2022.830260

Xu, S., & Xue, J. (2023). AI-driven transformation in the workplace: Role ambiguity, stress, and employee turnover intentions. *Technology in Society, 74*, 102312. https://doi.org/10.1016/j.techsoc.2023.102312

Yang, J., Zhang, Y., & Wu, H. (2020). Knowledge hiding and turnover intention: The mediating effect of organizational commitment. *Journal of Knowledge Management, 24*(6), 1443–1463. https://doi.org/10.1108/JKM-12-2018-0765

Yao, Z., Zhang, X., & Chen, L. (2020). The dark side of AI-induced job insecurity: The role of knowledge hiding. *Computers in Human Behavior, 105*, 106217. https://doi.org/10.1016/j.chb.2019.106217

Zhang, L., & Wang, H. (2020). The relationship between knowledge hiding and employee creativity: The moderating role of workplace trust. *Journal of Business Research, 114*, 380–389. https://doi.org/10.1016/j.jbusres.2019.09.037

Shikha Saloni is an Assistant Professor at ITM University specializing in Human Resource Management. Her research focuses on Workplace Incivility, Turnover Intention, Knowledge Hiding, and the Impact of AI on the Workforce. She has published in reputed journals and contributed to Scopus-indexed books. She actively teaches and mentors students in HRM and Organizational Behavior.

Neema Gupta is a Professor at Graphic Era Deemed to be University with over 20 years of experience in management education and research. She has guided PhD scholars, authored HRM textbooks, and edited international volumes. With numerous Scopus-indexed publications and academic awards, she continues to contribute significantly to research and academic leadership.

Enhancing Employee Data Management in Uganda's Private Sector: Challenges, Benefits, and Pathways to Responsible Practices

Mahadih Kyambade and Afulah Namatovu

Abstract With organizations becoming more and more dependent on digital systems, the responsible management of the data about employees has turned out to be a significant and important part of both ethical and legal standards. The chapter examines how the responsible employee data management practice has been adopted in the Ugandan companies in the private sector, its nature, scope, advantages, and obstacles. The study combines a literature review with empirical evidence gathered via case studies, semi-structured interviews, and surveys, which is why it is a mixed-method study. Thematic analysis showed that although a few organizations have implemented organized data protection, some organizations have issues with compliance, technological constraints, and resource constraints. The results provide some of the major advantages of responsible data management, such as more trust between the employer and the employees, better legal and regulatory adherence, and more operational efficiency. Nevertheless, there are still major hindrances that include a lack of awareness, a poor application of data protection regulations, and the high cost of installing an effective data protection system. The chapter highlights the importance of well-defined policies, educating employees, and enhanced regulatory control to guarantee that the use of employee data in the Ugandan private sector is carried out in an ethical and efficient way. The reinforcement of these points will not only improve the data security but also help to create a more responsible corporate culture, which will appreciate transparency and accountability. The study is relevant to the wider debate on data governance in emerging economies and will offer information to policymakers, corporate leaders, and regulators who need to enhance the practice of managing employee data in Uganda.

M. Kyambade (✉)
Department of Leadership and Governance, Makerere University Business School, Kampala, Uganda
e-mail: mkyambade@mubs.ac.ug

A. Namatovu
Department of Applied Computing & IT, Makerere University Business School, Kampala, Uganda
e-mail: anamatovu@mubs.ac.ug

M. Dzingirai and S. O. Atiku (eds.), *Strategic Human Resource Analytics and Intelligent Data-Driven Systems*, https://doi.org/10.1007/978-981-92-2623-8_4

Keywords Responsible data management · Employee data protection · Private sector · Uganda · Corporate governance

1 Introduction

In the modern era of the data-driven world, the issue of responsible data handling of employees has become the core focus of any organization in the world, including the accumulation of information concerning employees, their analysis, and protection as the key to organizational success (Stewart & Jurjens, 2017). The emergence of digital Human Resource Information System (HRIS), cloud computing, and data analytics has revolutionized the way information about employees is gathered, kept, examined, and used (Joshi & Sengupta, 2025). Firms worldwide are realizing that the poor handling of employee information not only affects trust and reputation negatively but also draws significant legal and regulatory repercussions (Bodie, 2022). In reaction, most jurisdictions, especially in Europe and North America, have established detailed data protection laws like the General Data Protection Regulation (GDPR), which require adherence to the letter and enforce accountability, consent, transparency, and data minimization in the treatment of employee data (Ussher-Eke, 2025).

The issue of responsible data management is still developing in Africa (Eke et al., 2022). Whereas certain countries, such as South Africa (with the Protection of Personal Information Act—POPIA) and Kenya (with the Data Protection Act of 2019), have made significant progress in making legal protection, enforcement is still uneven on the continent. Lack of technical infrastructure, a low level of enforcement of the existing regulations, and lack of awareness among both workers and employers make many entities in the private sector unable to align corporate data practices to the global standards (Klievink et al., 2016). Additionally, the expansion of digital systems in the African workplaces, in particular during and after the COVID-19 pandemic (Kyambade et al., 2023), has also led to increased concerns with data security, surveillance, and ethical handling of personal information on employees. These problems are particularly acute in Uganda. In spite of the fact that the country adopted the Data Protection and Privacy Act (DPPA) in 2019, it is still poorly enforced and the compliance of the private sector has been low. Numerous companies in Uganda still operate employee data in disjointed, old-fashioned, or insecure systems, and this makes sensitive personal information vulnerable to misuse, unauthorized access, or loss (Okori & Buteraba, 2024).

Employers in the private sector tend to have had little or no comprehensive data management policies, nor adequate training to be provided to the Human Resource (HR) staff as well as information technology personnel who are supposed to handle information belonging to employees. Furthermore, the level of internal responsibility is low, along with the lack of knowledge about employees on their right to personal information. Although there is an increasing adoption of digital HR systems, biometric attendance solutions, cloud-based payroll systems (Ammupriya et al., 2025), and e-performance management systems, scanty empirical data on the level at which the private sector firms in Uganda are incorporating responsible data practices into their HR governance models is known. Although the ethics, regulation, and effect of data governance in a corporate setting have been addressed by international and regional scholarship, the current study had a gap in research in the context of the Ugandan-specific corporate setting. Specifically, there is a shortage of empirical research examining the scope, difficulty, and prospects of responsible management of employee data. Most of the literature available focuses on the governmental institutions, and there is a gap in the knowledge of the way the private sector, which is comprised of an enormous and multiple diversity of businesses, manages these ever-increasing sensitive duties. As such, this research aims to describe and discuss how responsible employee data management practices are in the Ugandan private sector. The study will seek to analyze the advantages and difficulties to which the business can adopt responsible data governance and explore feasible ways towards more ethical, transparent, and effective employee data management. The overall aim of the project is to analyze how the responsible data management of employees in the private sector in Uganda can be implemented. In particular, the research is aimed at:

1. Determine the level and fruitfulness of employee data management in the organizations of the private sector.
2. Determine the advantages of having good data management such as increased compliance, operational effectiveness, and employee confidence.
3. Discuss barriers to the implementation of responsible data management, including technological barriers, resource barriers, and regulatory barriers.
4. Present practical suggestions to enhance employee data management practices, based on the best practices across the world and local factors.

Using the qualitative evidence, the study aims to educate the business leaders, policymakers, and regulators on the status of data practices, gaps that are currently present in the implementation, and interventions that are required to create a culture of accountability and data responsibility. Eventually, the research adds to the emerging data governance discourse in emerging economies and provides context-specific information to enhance corporate governance, safeguard the rights of employees, and support organizational integrity in the emerging digitalized economy of Uganda.

2 Literature Review

2.1 Management of Data about Employees: An International Outlook

In the digital economy, managing employee data has turned out to be a fundamental block of current human resource and governance frameworks. It also involves gathering, storing, usage, as well as protection of information about employees such as personal information, performance data, health data, and payroll data (Bodie, 2022). Companies all over the world are adopting responsible data practices in their operations as a way of complying with new data protection regulations like the European Union General Data Protection Regulation (GDPR), Californian Consumer Privacy Act (CCPA), or the Data Protection Act in the UK (ICO, 2022). The studies have shown that well-developed data governance models result in a better compliance level, lower legal risk, and a higher trust level of employees (Kyambade et al., 2025b; Sargiotis, 2024). Furthermore, employee information has become regarded as a strategic resource in the decision-making process, productivity analysis, and planning of the organization (Sousa et al., 2019). Nevertheless, responsible data management practices are still not uniformly used in most developing scenarios, as there are gaps in infrastructures, legal, and organizational contexts (Eke et al., 2022).

2.2 *Recent Trends in Employee Data Management in Africa and Uganda*

In Sub-Saharan Africa, data protection is becoming a trend, and most countries have introduced laws that seek to govern the utilization of personal information. In such countries as Nigeria, Kenya, and South Africa, a robust data protection policy has been established, which includes the Nigeria Data Protection Regulation (NDPR) and the Protection of Personal Information Act (POPIA) of South Africa (Modilim et al., 2024). Even with such efforts, it is still a problem in terms of enforcement because of the lack of institutional capacity and awareness of the issues by employers. The country of Uganda has also gone a long way with the introduction of the Data Protection and Privacy Act (2019) that requires data controllers and processors to take proper technical and organizational steps to protect data (Kyambade & Namatovu, 2025c). Nonetheless, it is still very weak and fragmented in the implementation of the same in the private sector (Nabukenya et al., 2023). According to a study conducted by Mukuki and Assenga (2024), although data protection policies have been incorporated by some big corporations in Uganda, most Small and Medium Enterprises (SMEs) continue to have basic measures of ensuring the security of the data of their employees. Such practices tend to put sensitive employee information at risk of unauthorized access, misuse, and loss.

2.3 *The Advantages of Good Employee Data Management*

Good data management of employees implies several organizational advantages. First, it improves practices of adhering to legal and ethical performance to protect organizations against litigation and reputational losses (Bayya, 2022; Kyambade & Namatovu, 2025b). Second, it can also lead to operational efficiency to the extent of making the record-keeping more efficient, less redundant, and automated HR processes (Claude, 2024). Third, it builds trust among staff and the management, particularly in the case when transparency in data and consent practices becomes a part of the organizational culture (Bayya, 2022). Companies that have put employee data governance at the forefront have expressed their satisfaction with employee engagement and retention in Uganda. Masood et al. (2023) have asserted that the higher the level of employee trust and cooperation, the better an organization is placed regarding its communication about how data about their employees is compiled and utilized. It is especially significant when referring to digital HR systems, working remotely, and hybrid models that require a strong flow of digital information.

2.4 Challenges to Responsible Management of Data in the Ugandan Private Sector

In as much as the benefits are clear, there are a number of challenges that do not support the successful implementation of responsible employee data management in Uganda. Most important of these is the lack of proper technological infrastructure; most organizations do not have secure digital platforms and use outdated and manual systems (Besigomwe, 2025). There is also the issue of having resource constraints, especially for small and medium enterprises, which may not have the capacity to purchase the secured software or data security officers. Moreover, the general lack of knowledge and awareness of the rights to privacy of data and its responsibilities among employers and employees is high (Kayondo & Kyobe, 2025). Although the regulatory environment is changing, it is rather under-enforced because of the lack of capacity in the Personal Data Protection Office (PDPO) that is in charge of compliance with the Data Protection and Privacy Act (Nabukenya et al., 2023).

2.5 Possible Pathways to Responsible Practices: International Best Practices and Domestic Adoptions

The international best practices offer effective guidelines on how employee data can be managed in Uganda. As an example, the GDPR model focuses on accountability, limiting the purpose, minimizing data, and the rights of data subjects (Torre et al., 2021). Such principles can help the Ugandan firms develop clear and ethical data practices. In the case of Uganda, it implies investing in HR professionals and IT personnel capacity building, creation of sector-specific guidelines, and building partnerships among the government, academia, and the private sector (Kyambade et al., 2024a, 2025a, b; Tumwebaze Alicon & Kalinaki, 2023). Uganda initiatives of digital transformation within the National Development Plan III (NDPIII) also offer possibilities of sector integrations to secure HR information systems (Kalema, 2024). The literature reviewed highlights the increasing significance of employee data management across the world as one of the pillars of corporate governance and compliance. Although the private sector in Uganda has demonstrated gradual improvements in using responsible data practices, some of the challenges that remain systemic in this country include poor infrastructure, lack of awareness, and poor enforcement of regulations. This research helps address the gap in the existing body of research on the importance of Ugandan business institutions in handling employee information by evaluating the existing practice, discussing the advantages and limitations, and offering evidence-based solutions to the problem that may be informed by both international standards and local context.

3 Methodology

The research design that is chosen in this study is a qualitative research design that is informed by an interpretivist paradigm to examine the application of responsible employee data management practices among the private sector in Uganda. A qualitative method is appropriate in revealing the depth, complexity, and situational aspects that contribute to the organization management of employee data (Creswell & Poth, 2016). The research focused on the thematic analysis in order to derive meaningful patterns from the participants and organizational practices.

3.1 Research Design and Rationale

Due to the exploratory nature of the set of research objectives, a multiple case study design has been adapted, which allowed a rich, contextual interpretation of data management practices in various organizations of the private sector. It is a design capable of in-depth investigation into real-life situations in which employee data are gathered, stored, and additionally processed to be able to draw comparative inferences and patterns (Yin, 2018).

3.2 Population of the Study and Sampling

The target population was human resource managers, IT officers, compliance officers, senior administrators of the selected organizations in the private sector that operate in Uganda. These people have been selected due to their direct participation in data management and strategic decision-making of employees. The participants were found using organizational directories and professional networks, and then official invitation letters and study information sheets were distributed by email to notify the participants about the study and ask them to participate willingly.

The study involved 12 organizations from the finance, telecommunications, health, manufacturing, and ICT sectors, which were selected using purposive sampling. In every organization, there were 3–5 key informants, making a total of 40 respondents. This sample can be considered large, though given that it was considered necessary to have the necessary representation in the sample, drawing a wide range of sectors and organizational levels, which will contribute to the richness, credibility, and transferability of the results. The involvement of various views of various functional departments made it possible to conduct more in-depth triangulation of the findings concerning the employee data management practices in the private sector of Uganda. The sample was also diverse in terms of organizational size, ownership (local and multinational), and degree of digitalization to give a wide scope of practices and issues.

The sample size was 40 respondents (23 men and 17 women), which represents the gender representation in managerial and technical employment in the Uganda private sector. The participants were mostly aged between 30 and 55 years old, with the greater proportion having a bachelor's degree or higher, with some of the senior administrators and compliance officers having a postgraduate degree in management, information systems, or human resource management. The respondents' professional experience, ranging from 5 to 20 years, represents a blend of mid-level and senior management positions across various departments, including Human Resources, ICT, Compliance, and Administration. This gender, age, education, and professional background diversity augmented the qualitative insights to obtain diverse views related to employee data management practices in different organizational settings.

3.3 Data Collection

Data were collected through semi-structured interviews, document analysis, and key informant consultations to ensure depth, credibility, and triangulation. Semi-structured interviews served as the primary method, offering flexibility to explore participants' experiences and perceptions in detail while maintaining consistency across the key thematic areas. Moreover, the internal organizational documents were also checked (employee data management policies, ICT guidelines, and compliance reports) to prove and supplement the information received during interviews. The consultations with key informants were conducted with the officials of the National Information Technology Authority—Uganda (NITA-U) and the Uganda Data Protection Office in order to involve expert opinions (Kyambade & Namatovu, 2025a). The interactions were insightful into the regulatory landscape and institutional support systems that could affect the data management behaviors. The interviews were held in the period between January and March 2025, and each session took about 45–60 minutes. All interviews were recorded on audio with the consent of the interviewees, and additional field notes were made to record non-verbal communication as well as background information.

3.4 Data Analysis

Thematic analysis was employed in the qualitative data analysis in this study in accordance with the six-step process suggested by Braun and Clarke (2006). To begin with, data familiarization was achieved by reading the interview transcripts and documents repeatedly to have a profound immersion and knowledge. Second, inductive and deductive codes were created during the first stage of coding with the assistance of four research objectives. This assisted in getting both new knowledge and pre-determined focuses of interest. Third, the thematic development involved the clustering of similar codes into the dominant themes and sub-themes expressing the salient patterns across the various case studies. These themes were subsequently checked and analyzed so as to ensure internal consistency, clarity, and high relevance to the research questions being targeted by the study. The fifth step involved defining and naming each of the themes, paying close attention to both the relationship with the existing literature and the unique findings revealing the Ugandan context of the private sector. Lastly, the themes were viewed by using the prism of the local and global data governance practices, providing a grounded but comparative insight. NVivo 14 software was applied to structure and code qualitative data and handle it in a systematic and traceable manner.

3.5 Trustworthiness and Rigor

In order to make the study more trustworthy and rigorous, a number of strategies were used to guarantee the credibility, transferability, dependability, and confirmability. Credibility was achieved by using member checking, where preliminary findings were discussed with a few participants in order to test their interpretation and also to make sure that their views were well represented. Triangulation was further utilized by incorporating the information from various sources such as semi-structured interviews, organizational documents, and expert consultations. This enabled the researchers to cross-verify the information and thereby enhance the consistency of developing themes. In order to ensure reliability, a detailed audit trail was kept over the course of the research process to record all the important decisions, methodological steps, and data analysis processes. This permitted transparency and possible imitation. Lastly, the issue of confirmability and transferability was taken care of through employing the thick contextualized description in the presentation of findings. These plentiful stories allow the readers and researchers to evaluate the extent to which the findings can be applied in other similar situations, particularly in developing countries where the issue of data governance in the private sector is becoming increasingly pertinent.

3.6 Ethical Considerations

Informed consent was obtained from all participants, and all data were handled in accordance with Uganda's Data Protection and Privacy Act (2019). There was assurance of anonymity and confidentiality. The codes were given to organizations and individuals to ensure the protection of their identities. All data were properly stored and were then destroyed when the study was completed in accordance with ethical standards.

4 Findings

This study has found that the findings are structured thematically in terms of the four research objectives. Interpretation of the interview transcripts and organization documents of 12 organizations in the private sector in Uganda was done using NVivo 14. The analysis came up with four themes and a number of sub-themes, as shown below. Themes that surfaced included (1) existing data management process in employees, (2) advantages of good data management, (3) implementation barriers, and (4) suggested improvement strategies. Under each theme, we provide both aggregate results and differences between different chosen case study organizations (anonymous as Org A-L). The participants comprised the HR managers, IT officers, compliance officers, and top-level managers of organizations operating in the banking, manufacturing, telecom, hospitality, and education industries.

4.1 Theme 1: Recent Practices and Level of Employee Data Management

The management practices at both ends of employee data management were quite diverse in organizations and mostly depended on the size of the organization, the field, and the technological ability.

Digital vs. Manual Systems: Larger organizations, especially multinational ones, have shifted to comprehensive Human Resource Information System (HRIS) and cloud-based systems. These systems were capable of updating employee records in real-time, automating payroll, and providing secure access.

> All of our HR operations are digitalized. Every employee will possess a profile with work history, management reports, and training history. (Org A, HR Manager)

Conversely, most Small and Medium Enterprises (SMEs) continue to use paper-based documents or primitive Excel-based systems, which are therefore prone to data loss and unauthorized access.

Data types obtained: Data obtained by the organization usually include personal identification information, education history, payroll history, next of kin, and health information (in certain instances) and biometric information (fingerprints used to check in and out). Nevertheless, inconsistencies in the classification and management of sensitive vs. non-sensitive data were also noted.

Data Access and Security Controls: Not many organizations had defined data access policies. In other instances, access to HR data was not regulated by various departments, and the chances of misuse were increased.

> It has no official hierarchy of access. Employee files can be accessed by any supervisor who has an employee company login. (Org E, IT Officer)

The majority of the organizations store the employee demographic and employment records on the digital HR systems (e.g., ERP, payroll software). Nevertheless, paper-based files with digital records were still present in Org D, Org G, and Org J.

> We have an internal HRMS, but the files are stored in a manual form as the backup of legal records. (Interviewee P03, Org D, HR Officer)

Organizations were very varied in their data maturity. Multinationals (Org A, Org F, Org L) were more sophisticated in encryption, biometric authentication, and data audits than local SMEs (Org B, Org E).

4.2 Theme 2: The Advantages of a Properly Managed Employee Data

Some of the advantages mentioned by participants in organizations that had implemented more organized data management systems.

Legal and Regulatory Compliance: Firms that were compliant with the Uganda Data Protection and Privacy Act (2019) and matched the global standard of data protection, for example; GDPR, were reported to have smoother audits and stronger reputational credibility.

> Compliance has turned out to be a competitive advantage. The international contracts we will be bidding on now will no longer be international, as our data processes will be in line with international expectations. (Org F, Compliance Manager)

Operational Efficiency: Digitized systems enhanced the HR operations through reducing mistakes in payroll processing, automated leave monitoring, and workforce planning.

Better Trust of Employees: Transparent data operations and explanation of data use created the feeling of safety among employees, which increased their morale and interest.

> When they realize that the information is secured, our staff becomes more transparent to share it. It builds trust. (Org C, HR Director)

> Correct computer records have assisted us in the simplification of payroll and performance tracking. Reports are also easier to create for URA and NSSF. (P11, Org C, Interviewee, IT Manager)

Org L (which is ISO certified), Org A, and Org F recorded that the data management became structured and thus improved their eligibility to partnerships and audits. Conversely, Org I and Org K stated that they had a problem complying with the Data Protection and Privacy Act of Uganda because of the lack of capacity.

4.3 *Theme 3: Obstacles to Responsible Data Management*

Although there was recognition of the importance of data management, most organizations struggled to implement it due to several interconnected challenges.

Technological Limitations: Low ICT infrastructure, old software, and intermittent internet connection were the problems experienced, particularly in the case of SMEs and regionally-based organizations.

Low Awareness and Capacity Gaps: The general level of awareness about data protection principles was low both at the executive and the staff level. Some employees had not been trained properly on how to handle data.

> The Data Protection Act does not fully understand what the Data Protection Act entails even on the part of the management. It is regarded as a burden of compliance rather than business focus. (Org G, Legal Officer)

Weaknesses in Regulatory and Policy: A number of firms did not have any policy on internal data protection or their policies did not work well. Internal accountability was undermined further because of the lack of specific Data Protection Officers (DPOs).

Financial Constraints: The financial constraints hampered the investments in powerful HRIS systems, employee training, and cybersecurity mechanisms.

> The policy of data protection is not known to most of our staff. Even the top management believes that it is a Western problem. (P17, Org H)

Professionals who participated in the study's interviews stated that they received 3 months of training on this aspect before assuming their duties. According to the interviewee who played the role of compliance officer in the study, 3 months of training in this area were provided before they were given their duties. Org B, Org D, and Org G mentioned that they were hindered by unstable internet connections and a lack of IT expertise. Some organizations did not have specific officers in charge of data protection or internal regulations that would be in line with the national law.

4.4 *Theme 4: Suggestions and Improvement Solutions*

Participants recommended a number of measures that can be used to enhance responsible employee management practices in the Ugandan private sector.

Capacity Building and Awareness Campaigns: At the request of the internal (employee-level) and external (industry-wide) training as a means to create knowledge and change the attitude towards data privacy.

> Training must go beyond HR. Every data mishandler must have the knowledge of what is at stake. (Org H, Training Coordinator)

Internal Policy Formulation and Implementation: Organizations reported the desire to have standard templates and industry policy manuals. It was considered important to have a dedicated role/committee for data governance.

Investment in Technology: Respondents suggested that the government should incentivize or collaborate with the HRIS solutions and cloud storage options to make them more affordable to SMEs.

Regulatory Support/Sectoral Benchmarking: Firms requested proactive involvement on the part of the National Information Technology Authority-Uganda (NITA-U) and industry associations to give advice and facilitate best practices. It would help to relieve the burden on small players by making regulators more proactive, e.g. by using sector-specific toolkits or sample policies. (Org K, CEO) NITA-U and sector regulators need to be more practical. We are currently in the process of making guesses when it comes to compliance. - P30, Organization, J, Human Resources Manager. The members of Org F and Org A suggested the inclusion of Data Protection Impact Assessments (DPIAs) and staff sensitization workshops. SMEs demanded the subsidization of ICT tools and industry-specific templates to facilitate policy making. Table 4.1 shows the cross-case analysis matrix.

Table 4.1 Cross-case analysis matrix

Theme	Org A (multinational)	Org B (local SME)	Org F (multinational)	Org G (local NGO)	Similarities
Present practices	Advanced HRMS with encryption	Hybrid system	Biometric and cloud	Manual records	The majority of organizations collected employees' biodata and salary information
Benefits	Better compliance and efficiency	Fewer payroll errors made	Regulatory recognition	Minimal employee trust	Tech improved trust
Challenges	Upgrading systems cost	Lack of awareness and skills	Cybersecurity threats	No policy framework	Regulatory gaps and training needs
Recommendations	Global benchmarking	Subsidized tools	Hire data protection officer	Internal policies	Training recommendations

5 Discussion

The research was aimed at investigating the application of responsible employee data management in the private sector in the Republic of Uganda. The findings are presented under the four objectives, and the landscape displayed in this case is both promising and challenging to a significant degree. These findings were then interpreted in the following discussion in terms of the existing literature and theoretical frameworks with implications for policy, practice, and future research.

5.1 *Employee Data Management Practices: Lack of Balanced Development and Structural Constraints*

The results indicate that there is a high level of difference between the employee data management practices of the large corporations and Small to Medium Enterprises (SMEs) in Uganda. The bigger organizations, particularly those that have international associations, have embraced digital HR systems and structured systems of data governance. The vast majority of SMEs, conversely, continue to use paper-based or semi-automated systems (Kayanja et al., 2025; Kyambade et al., 2025c), which makes them more susceptible to data breaches and inefficiencies. This is consistent with the previous literature that indicates that digitization of HR functions is usually not uniformly applied in developing economies since they lack equal resources and technological potential (Mukhuty et al., 2022; Nosike, 2024). As an example, Oduro (2020) established that SMEs in Ghana had comparable limitations in the adoption of structured data systems because of insufficient funds and technical expertise. The absence of uniform policies of employee data protection also rings with the results of Eke et al. (2022), who emphasized the fragmentation of policies and the contrived institutional capacity as the main obstacles to data governance within Southern Africa. These trends indicate that the digital divide in HR management is not only technological in Uganda but also organizational and regulatory.

5.2 *Responsible Data Management Benefits: Trust, Efficiency, and Compliance*

The research determined that the employed practices of responsible employee data introduced increased the efficiency of the organizations, the development of employee trust, and legal compliance. These advantages are a reflection of the propositions in world literature. Sousa et al. (2019) argue that responsible data stewardship companies have more workforce engagement and better reputational capital. This is again affirmed by the Ugandan case. The participants of the companies that had good systems engaged had lower payroll error, enhanced internal audit performance, and easier compliance with the Uganda Data Protection and Privacy Act (2019). These results find parallels with Eke et al. (2022) and Namatovu and Kyambade (2025a) who discovered that in African businesses, the control of employee data helped to establish a more responsible and law-abiding corporate culture. It should be noted that improved employee trust was one of the themes. This is in line with the argument of Liao (2022), who stated that open data practices enhance the strength of psychological contracts between employers and employees, particularly in places with a traditionally low level of labor protection.

5.3 Impediments to Implementation: A Triad of Technology, Awareness, and Regulation

Although the role of responsible data practices is acknowledged, a number of obstacles remain, most prominently technological reasons, lack of knowledge, and ineffectiveness in enforcing regulations (Kyambade et al., 2024b). The results confirm the body of literature that highlights structural challenges to digital transformation in sub-Saharan Africa (Achieng & Malatji, 2022; Namatovu & Kyambade, 2025b). This research states the low priority of data governance at the executive levels as a constraint. Although the senior managers might be aware of the strategic importance of digital HR systems, the lack of continuity in training and capacity building compromises the willingness of the institutions to manage the data in a responsible manner. The observation of the financial constraints hindering the adoption of HRIS in SMEs reflects previous studies by Kenetey and Popesko (2025), which found SMEs in Ghana to be faced with the same budget constraints. Furthermore, the absence of an appointed Data Protection Officer within the majority of companies indicates a serious compliance breach, as mentioned by Lancieri (2022).

5.4 Paths to Responsible Data Management: Contextualized and Collaborative

The respondents proposed effective avenues of implementing employee data management that could be taken in the areas of specialized employee training, standardization of policies, and regulation. These suggestions resonate with the suggestions of the literature that urges the locally tailored and capacity-oriented approaches in data governance (Eke et al., 2022). The recommendations of the increased cooperation among the private sector, government (especially NITA-U), and professional organizations indicate that the recommendations of Sanner et al. (2025) are relevant because they recommend ecosystem-wide partnerships to enhance ICT governance in Uganda. This study addresses the long-standing issue in the African setting related to the affordable and scalable nature of digital solutions by promoting scalable and cost-effective HR technologies, including open-source solutions and cloud-based systems (Eke et al., 2022; Namatovu & Kyambade, 2025c).

In addition to the technical and regulatory aspects, the coordination of employee data management with the cultural and ethical background of Africa is also essential to attain social legitimacy as well as moral responsibility. In Uganda and a great part of sub-Saharan Africa, the values of Ubuntu, which is a philosophy of human dignity, community, and mutual respect, should be the foundation of data practices, as opposed to the individualistic orientation of most models of data governance in the West. Adopting Ubuntu principles means that organizations do not simply consider employee data as an efficiency tool, but as part of the human identity that is to be treated, kept secret, and held accountable as a community. This, in practice, entails the development of participatory data governance practices, in which employees are significantly informed and consulted on the means of collecting, storing, and using their data. Culturally responsive frameworks of this kind promote openness, promote organizational trust, and reduce the chances of infringing on primary social principles. Combining aboriginal ethics with legal standards, including the Data Protection and Privacy Act of 2019 of Uganda, companies will be able to establish a middle ground of responsible data handling that is both internationally acceptable and nationally lawful.

6 Theoretical Implications

The study contributes to the theoretical discussion of responsible data governance as it puts the applicability into the context of the human resource management of low-resource and developing environments, namely the private sector of Uganda. It provides empirical evidence of the Institutional Theory in operation (Scott, 2001) and provides an example of how the regulatory (legal requirements, such as Uganda Data Protection and Privacy Act, 2019), normative (industrial expectations and HRM standards), and cognitive (internal values and leadership commitment) forces interact to influence organizational behaviors in the context of data management of employees. It is interesting to note that the researcher states that multinational corporations (MNCs) are more prone to formal data governance practices since they are more exposed to outside regulatory compliance concerns, both international and local. Small and Medium-Sized Enterprises (SMEs), on the contrary, tend to be exposed to less institutional pressure and less control, and therefore have fragmented or informal practices.

The results also support the Technology-Organization-Environment (TOE) Framework (Tornatzky & Fleischer, 1990), which supports the adoption of technological innovation depending on three contextual factors, namely, technological readiness, organizational capabilities, and environmental influences. In this context, the research demonstrates that organizational capacity (e.g, skilled staff and IT infrastructure), technological complexity (e.g., issues about cost, integration, and digital literacy), and external uncertainties (e.g., weak enforcement, changing regulations) have a direct influence on the degree and efficiency of employee data management practices. These constraints become major obstacles to responsible data governance, especially in SMEs. This research offers an in-depth insight into the interaction between institutional and structural variables and organizational behavior by applying and applying these theories in the environment of the developing economy. It requires a more adaptive application of these theories to meet social-economic and technological reality in the emerging markets. This input provides a basis on which to continue theorizing in digital ethics, data security, and HRM within such a low-resource context in the future.

7 Practical Implications

1. With respect to Private Sector Firms:

   This research highlights the importance of companies to consider the management of their employee data as a strategic organizational activity and not an administrative routine. Data governance by responsible individuals will help increase the efficiency of operations, reduce legal risks, and increase the trust that employees have. The recommended approach to cost-effective HR systems ought to be adopted by organizations, particularly SMEs, and periodic training of HR staff and IT employees on data ethics, privacy, and security measures. Best practices can also be further institutionalized by setting internal data protection policies and creating special data protection officers.
2. To the Policymakers and Regulators:

   Regulatory authorities, including the National Information Technology Authority—Uganda (NITA-U) and the Personal Data Protection Office (PDPO), need to step up the enforcement process and help ensure compliance by providing advice and building capacity. This also involves coming up with sensible toolkits, compliance templates, and sector-related data protection standards. Frequent interaction with the private sector in the form of workshops, audits, and consultancy services would assist in the reduction of gaps in implementation and increased regulatory compliance, especially by resource-constrained SMEs.
3. In the Case of Educational and Training Institutions:

   There is an acute necessity for academic institutions and professional training to implement academic modules on data governance and cybersecurity, and digital ethics in their programs, particularly in the fields of HRM, IT, and business management. Providing the future workforce with the foundational knowledge and practical skills on employee data protection will create a more prepared workforce that can address any emergent digital and ethical issues in the work environment.

4. In the Case of Civil Society and Industry Associations:

   Associations like chambers of commerce, professional associations, and civic society organizations can be at the forefront in promoting responsible data practices, and also as a platform to learn with peers. These actors can contribute to the normalization of ethical employee data management as the pillar of contemporary corporate governance by way of organizing sectoral discussions and sharing knowledge.

8 Limitations and Future Research Directions

The research is highly informative on the issues of responsible employee data management in the private sector of the Ugandan economy, though it was limited in multiple aspects. The multiple case study design that is qualitative and rich in detail restricts the generalization of results to the whole population of organizations sampled. The differences in size, industry, and location may influence data practices not addressed in this study. The sample size was limited and could not be representative of the entire scope of practices among sectors, especially those that were less represented and informal ventures.

The data were self-reported mostly by the HR managers and were subject to social desirability bias, particularly in matters relating to law. Also, the researchers did not critically examine the views of major regulatory agencies, and the actors of the private sector were considered, which significantly restricted the understanding of policy implementation. It was also a cross-sectional study, and hence it was not able to follow changes over time, which is critical in a quickly changing digital and legal context. Quantitative or mixed methods designs should be taken into consideration in future studies to conduct a wider generalization and further analysis. Sectoral studies can be done comparatively to identify industry-specific practices, and longitudinal research may be conducted to trace the progress with respect to the changing regulations. Incorporating employee feedback would bring out more ethical and trust-related issues. Finally, the research based on regulators and other stakeholders would contribute to the knowledge of the data governance situation in Uganda and contribute to the creation of more efficient policy and organizational strategies.

9 Conclusion

The chapter highlights the need to adopt responsible management of employee data in the private sector in Uganda. Some organizations are mature in this respect, but there still exist systemic barriers. To address these gaps, it will be necessary to use a multi-pronged approach that simultaneously takes into account technology, awareness, and regulation. The research can help understand the ways to localize data governance principles to the developing situation better, and the research and practice can be based on this study in the future.

References

Achieng, M. S., & Malatji, M. (2022). Digital transformation of small and medium enterprises in sub-Saharan Africa: A scoping review. *The Journal for Transdisciplinary Research in Southern Africa, 18*(1), 1257.

Ammupriya, A., Vaishnavi, S., Ashwini, A., Kavitha, R., Paranthaman, P., & Saravanan, V. (2025, March). *Cloud-based HR platforms for scalable workforce Management in Multinational Organizations*. In 2025 3rd international conference on disruptive technologies (ICDT) (pp. 1607–1613). IEEE.

Bayya, A. K. (2022). Advocating ethical data management and security. *International Journal of Scientific Research in Computer Science, Engineering and Information Technology (IJSRCSEIT),8*, 396–417.

Besigomwe, K. (2025). The impact of digital platforms on taxation in Uganda: Challenges and opportunities for the Uganda revenue authority. *Cognizance Journal of Multidisciplinary Studies, 5*(1), 85–104.

Bodie, M. T. (2022). The law of employee data: Privacy, property, governance. *Indiana Law Journal, 97*, 707.

Braun, V., & Clarke, V. (2006). Using thematic analysis in psychology. *Qualitative Research in Psychology, 3*(2), 77–101.

Claude, K. J. (2024). The impact of records management on the effectiveness of human resource Management in the Public Sector. *International Journal of Research and Innovation in Social Science, 8*(5), 1040–1060.

Creswell, J. W., & Poth, C. N. (2016). *Qualitative inquiry and research design: Choosing among five approaches*. Sage publications.

Eke, D., Ochang, P., Adimula, A., Borokini, F., Akintoye, S., Oloyede, R., … & Ogundele, T. (2022, September 4). *Responsible Data Governance in Africa: Institutional gaps and capacity needs*. Centre for the Study of African Economies (CSEA).

ICO. (2022). *UK data protection act 2018: Guide to data protection*. Information Commissioner's Office. https://ico.org.uk/for-organisations/guide-to-data-protection/

Joshi, K. A., & Sengupta, S. (2025). Information systems in HR processes: Perspectives and processes. In *Research handbook on information systems and technology* (pp. 191–209). Edward Elgar Publishing.

Kalema, N. L. (2024). The 'digital transformation for Development' Anti-politics machine: A case study on global digital development governance and public-sector digital transformation in Uganda. *Policy & Internet, 16*(4), 750–763.

Kayanja, W., Kyambade, M., & Kiggundu, T. (2025). Exploring digital transformation in higher education setting: The shift to fully automated and paperless systems. *Cogent Education, 12*(1), 2489800.

Kayondo, B. N., & Kyobe, M. (2025). The influence of self-efficacy, trust, perceived costs and healthcare-environment on compliance with the data protection and privacy act by health Workers in Uganda. *The Electronic Journal of Information Systems in Developing Countries, 91*(2), e70002.

Kenetey, G., & Popesko, B. (2025). Budgetary control and the adoption of consortium blockchain monitoring system in the Ghanaian local government. *International Journal of Public Sector Management, 38*(1), 12–29.

Klievink, B., Bharosa, N., & Tan, Y. H. (2016). The collaborative realization of public values and business goals: Governance and infrastructure of public–private information platforms. *Government Information Quarterly, 33*(1), 67–79.

Kyambade, M., & Namatovu, A. (2025a). Health-care leaders' perspectives on AI implementation in Uganda: Overcoming barriers, driving innovation and strategic considerations. *Leadership in Health Services, 38*(3), 442–463.

Kyambade, M., & Namatovu, A. (2025b). The impact of ethical leadership and code of ethics on employee integrity, mediated by job satisfaction in Uganda's public sector. *International Journal of Public Leadership, 21*(1), 72–89.

Kyambade, M., & Namatovu, A. (2025c). Transforming the public service via digital leadership in Uganda. *International Journal of Public Leadership, 21*(1), 37–53.

Kyambade, M., Abdul, S. M., & Namatovu, A. (2023). Adoption theories for internet banking in Uganda. *Advances in Research, 24*(6), 136–145.

Kyambade, M., Bartazary, D., & Namatovu, A. (2024a). The impact of motivation on civil servants' performance in Tanzania. *Journal of Management Research, 24*(1), 11–18.

Kyambade, M., Kagere, R., Namatovu, A., Ssentumbwe, A. M., Tushabe, M., & Oyella, S. (2024b). Technological innovation and commercial banks' financial performance: A mediation analysis of risk management practices. *SEISENSE Business Review, 4*(1), 98–117.

Kyambade, M., Kisseka, R., & Namatovu, A. (2025a). Blended learning models in Ugandan higher business education: Best practices and lessons learned. *Cogent Education, 12*(1), 2545329.

Kyambade, M., Namatovu, A., & Male Ssentumbwe, A. (2025b). Exploring the evolution of artificial intelligence in education: From AI-guided learning to learner-personalized paradigms. *Cogent Education, 12*(1), 2505297.

Kyambade, M., Namatovu, A., & Tushabe, M. (2025c). Transformational leadership, corporate governance practices and small and medium-sized enterprises (SMEs) performance: A developing nation perspective. *SN Business & Economics, 5*(9), 1–26.

Lancieri, F. (2022). Narrowing data protection's enforcement gap. *Maine Law Review, 74*, 15.

Liao, Y. (2022). How does the labor protection law affect sustainable economic growth: An empirical analysis based on psychological contract perspective. *Frontiers in Psychology, 13*, 928282.

Masood, A., Zhang, Q., Ali, M., Cappiello, G., & Dhir, A. (2023). Linking enterprise social media use, trust and knowledge sharing: Paradoxical roles of communication transparency and personal blogging. *Journal of Knowledge Management, 27*(4), 1056–1085.

Modilim, S. N., Bolarinwa, I. Y. A. N. U. O. L. U. W. A., Omidiora, M. T., & Tawo, O. B. A. H. (2024). Reforming data governance in Nigeria: A critical analysis of the Nigeria data protection act, regulatory enforcement, and global alignment. *International Journal Of Law Management & Humanities, 7*(6), 2481–2495.

Mukhuty, S., Upadhyay, A., & Rothwell, H. (2022). Strategic sustainable development of industry 4.0 through the lens of social responsibility: The role of human resource practices. *Business Strategy and the Environment, 31*(5), 2068–2081.

Mukuki, A., & Assenga, A. (2024). *Comparative study of data protection legislation frameworks across the east African community*. D4D Hub, March, 2024–2005.

Nabukenya, J., Drumright, L., Alunyu, A. E., & Semwanga, A. R. (2023). Critical risk and success factors for sustainability of an electronic health data capture, processing and dissemination platform for Uganda. *Health Informatics Journal, 29*(2), 14604582231180576.

Namatovu, A., & Kyambade, M. (2025a). Artificial intelligence and employee performance in Uganda's healthcare institutions: Exploring the mediation effects of perceived ease of use and skills enhancement. *Journal of Health Organization and Management*.

Namatovu, A., & Kyambade, M. (2025b). Assessing the impact of digital leadership on public sector performance: The mediation role of digital transformation in developing economies. *SAGE Open, 15*(3), 21582440251367585.

Namatovu, A., & Kyambade, M. (2025c). Leveraging AI in academia: University students' adoption of ChatGPT for writing coursework (take home) assignments through the lens of UTAUT2. *Cogent Education, 12*(1), 2485522.

Nosike, C. J. (2024). Digitalization in developing countries: Opportunities and challenges. *Nigerian Journal of Arts and Humanities, 4*(1), 2955-0343.

Oduro, S. (2020). Exploring the barriers to SMEs' open innovation adoption in Ghana: A mixed research approach. *International Journal of Innovation Science, 12*(1), 21–51.

Okori, W., & Buteraba, S. (2024). Cyber security exploits and management in telecommunication companies: The case of Uganda. *Journal of Computer Science and Technology Studies, 6*(4), 67–76.

Sanner, T. A., Kempton, A. M., Russpatrick, S., & Sæbø, J. I. (2025). Governing digital platform ecosystems for social options. *Information Systems Journal, 35*(2), 422–449.

Sargiotis, D. (2024). Key principles of data governance: Building a strong foundation. In *Data Governance: A Guide* (pp. 137–163). Cham.

Scott, W. R. (2001). *Institutions and organizations* (2nd ed). Sage.

Sousa, M. J., Pesqueira, A. M., Lemos, C., Sousa, M., & Rocha, Á. (2019). Decision-making based on big data analytics for people management in healthcare organizations. *Journal of Medical Systems, 43*(9), 290.

Stewart, H., & Jürjens, J. (2017). Information security management and the human aspect in organizations. *Information & Computer Security, 25*(5), 494–534.

Tornatzky, L., & Fleischer, M. (1990). *The process of technology innovation*. Lexington Books.

Torre, D., Alferez, M., Soltana, G., Sabetzadeh, M., & Briand, L. (2021). Modeling data protection and privacy: Application and experience with GDPR. *Software and Systems Modeling, 20*(6), 2071–2087.

Tumwebaze Alicon, A., & Kalinaki, K. (2023). Positioning higher education institutions as work-based ICT-integrated learning theatres for employee mid-career development; a strategy for HR capacity building. *Higher Education, Skills and Work-Based Learning, 13*(5), 955–968.

Ussher-Eke, D. (2025). HR and GDPR: Partnering to protect employee data. *World Journal of Advanced Research and Reviews, 27*(2), 717–730.

Yin, R. K. (2018). *Case study research and applications* (Vol. 6). Sage.

Mahadih Kyambade is a distinguished Academician and Lecturer in the Department of Leadership and Governance at Makerere University Business School. Holds a PhD in Responsible Leadership, Master of Science in Leadership and Governance, Bachelor of Leadership and Governance and Bachelor of Laws from Makerere University. His research focuses on themes related to Leadership, Governance, Psychological Safety, Gender Diversity, Higher Education Management, Sustainability performance, Environmental Studies and Policy.

Afulah Namatovu is a renowned Academician in the field of Information Management. She is a Lecturer in the Department of Applied Computing & IT, Makerere University Business School. Holds a Bachelor's degree in Information and Information Management from Makerere University. Her research focuses on Information Management, IT, Generative AI, Leadership and Gender Issues in Organizations.

Leveraging Personalized Employee Development Plans to Support Change Management Models in Academic Libraries in Zimbabwe

Janeth Matingwena and Admore Mashokoh

Abstract This chapter explores how personalized employee development plans assist academic libraries in Zimbabwe in implementing change management models. As the pace of technological change in academic libraries accelerates and the librarian is required to pivot to new tools, services and systems at a rapid pace, the need arose for the research. The data were gathered through semi-structured interviews with 20 librarians sampled purposively from 10 state universities. The study also used Lewin's Change Management Model. The study found that most academic libraries did not have custom employee development plans to support their change process. The study further stated that the potential value of custom employee development plans could support academic libraries in training their librarians with the skills needed to enable them to manage change. The study suggested that the personalized employee development plans supported strategic planning, communication, and collaboration. The study concluded that the academic libraries in Zimbabwe would benefit from using custom employee development plans to support change management models. The study recommended that the academic libraries should embrace custom staff development plans to enhance librarians' capacity for managing change and embed them into their change management component. Study findings recommended that the academic library management should offer adequate training, so librarians can earn the knowledge and skills required for effectively managing change. Lastly, the study recommended that, to ensure the successful implementation of custom employee development plans for support and demonstrate involvement as a collective, academic librarians should ensure that all relevant stakeholders are involved.

J. Matingwena (✉)
Department of Human Resource Management, Manicaland University of Applied Sciences, Mutare, Zimbabwe
e-mail: janeth.matingwena@staff.msuas.ac.zw

A. Mashokoh
Department of Scholarly Communication (Research Discovery and Dissemination, Great Zimbabwe University, Masvingo, Zimbabwe
e-mail: amashokoh@gzu.ac.zw

M. Dzingirai and S. O. Atiku (eds.), *Strategic Human Resource Analytics and Intelligent Data-Driven Systems*, https://doi.org/10.1007/978-981-92-2623-8_5

Keywords Academic Libraries · Change Management · Development Plans · Employee · Models

1 Introduction

Globally, academic libraries must implement employee development plans in order to meet the ever-changing library operations and services due to advancements in technologies (Tait et al., 2016). In this regard, Personalized Employee Development Plans (PEDPs) emerge as vital tools to generate organizational agility and employee tenacity. Saul et al. (2024) defined employee development plans as plans tailored to development for professional skills and addressing the specific developmental needs of an organization. Vezi et al. (2024) support this view and maintain that such plans must be in line with the objectives of an institution that aims to develop a culture of lifelong learning and proactive adaptation. Ghedabna et al. (2024) also argue that individual employee development plans are strategic devices to identify employee proficiencies and training needs, addressing the alignment of the interest of personal development with organizational requirements (Ellikkal & Rajamohan,2025). Yet, individuals are challenged in recognizing their path of personal employability development, when they do not have a personal employment development plan (Platt, 2020.

Large companies like Google, through its Project Oxygen, rolled out individual work plans for managers; the move lifted talk between teams, boosted staff happiness, and pulled people into their jobs (Mizrak, 2023). These illustrations demonstrate how individualized plans can inspire a more dedicated and effective workforce. Uka and Prendi (2021) add that when staff believe the company sees them and listens, output can soar. Crowley (2025) also remarks that firms that shape the work experience around the person notice stronger engagement and fewer people walking out the door. Sripada (2024) argues that using tech to weave this level of personalization into daily life has become a must in a fast-moving workplace. Therefore, to draw in and keep the best talent, employers now lean on custom training, one-on-one growth paths, and personal channels so each employee feels heard, respected, and ready to give their best.

Akuffo-Aduamah (2025) argues that tailoring employee development plans for each employee goes well past boosting engagement; it is, in fact, a strategic move that ties directly to the larger aims of the business. Kalandarovna and Qizi (2023) add that companies which invest in such personalized learning paths set themselves apart in a fast-changing market as long as technology keeps advancing. Babatope (2025) showed that plans work best when they clearly match a person's personal goals, can be used right away at work, and include regular chances for reflection or check-ins. The study also revealed common threads between traditional development schemes and upward leadership coaching, suggesting that programs aimed at skill building can, in a volatile, uncertain, complex, and ambiguous world, stretch an individual's competence and capacity. Collectively, these findings offer practical guidance on shaping personal growth plans into powerful tools for navigating change and uncertainty. Therefore, by investing in tools that facilitate tailored experiences, employers can cultivate a workforce that is not only more motivated but also more committed to achieving the organization's objectives. In this approach, the marriage of personalization employee development plans with technology becomes a driving force behind a resilient and adaptive workforce equipped to navigate the challenges of the future.

Saul et al. (2023) sought to understand why, between 2015 and 2020, the personal growth plans of South African Social Security Agency (SASSA) employees based in Gauteng deviated from the organization's larger training budget. According to their research, this discrepancy resulted in underwhelming employee performance and unused learning funds, which further reduced the value of scheduled workshops and courses. The team proposed a set of doable steps that SASSA could implement in the 2024–2025 fiscal year to improve the situation and align the agency's development initiatives with the objectives of its employees and the broader strategic requirements. Among these is a push for digital learning paths and newer training models grounded in solid theory, all aimed at boosting individual and organizational output. Building on that, Swanson (2022) turned to a more open question, studying employee self-improvement as part of Human Resource Development and looking at it through a network lens that was seldom used. The study found that individuals aim to further their personal development by taking personal responsibility for learning and development.

Research shows that pulling off real change is tough, which partly explains why so many organizations struggle to get it right (Errida & Lotfi, 2021). Every planned change affects all employees, either through noticeable disruptions or small, gradual adjustments in their daily work activities (Burke, 2023). To avoid confusion, a clear personal development plan must link a person's learning goals to the achievement of the organizational biggest aims and objectives. George et al. (2025) add that the plan works best when it mixes aspirations at home with targets on the job, giving growth a full-time shape. Fredberg and Pregmark (2022) argue that, without ongoing change, even established institutions like academic libraries can slowly drift toward redundancy. They found that steering change well is vital when new technologies knock at the door. The authors also noted that scholars and managers alike have built a toolbox of change models in order to construct a shared language about what change really means. Chychun et al. (2023) assert that change management models serve as a guide to assist in navigating change initiatives. They essentially identify the key levers that drive success in the change management journey by highlighting the different factors that can impact change and pinpointing specific processes. Several models have been developed, based on a variety of theories and concepts from various disciplines (Phillips & Klein, 2023). It is critical to understand that there is not a single ideal change management strategy or model. The real situation and the possibility of combining several models should be considered when choosing a specific approach (Rachmad, 2022). As such, using only one or a small number of models is insufficient to handle the variety of situations that change may bring (Rožman et al., 2023).

A key competency that influences an institution's development and aids in its accomplishment of objectives is effective change management (Rožman et al., 2023). Resistance frequently results when organizational changes do not align with what staff members already know and are capable of. By ensuring that employees have the necessary skills to adjust to new roles and responsibilities, customized employee development plans can play a crucial role in addressing this issue and reducing resistance to change. According to Aslam et al. (2025), library administrators must have a clear vision, a strategic plan, self-assurance, and the capacity to collaborate with a variety of groups in order to meet future challenges. Additionally, Rahman et al. (2025) stress that library workers need to be adaptable and willing to learn new skills and adopt fresh approaches to interacting with the academic community. For example, Oyedokun (2025) investigated how academic libraries can facilitate library system switching. This study investigated how academic libraries can manage changes more smoothly, highlighting the importance of having a good change management approach. Kurt Lewin's theory, ADKAR, Kotter's model, and the McKinsey 7s framework were the four main models it drew from. The results demonstrated that many libraries have trouble managing change because they frequently lack the infrastructure and specialized resources needed to transition their library systems, as well as knowledge about how to execute these changes successfully.

One of the main causes of the changes occurring in academic libraries is technological advancements. Weinper and Pietrzyk-Leonowicz (2022) state that libraries are concentrating on making sure they have the newest technology, improving search capabilities, facilitating access to electronic resources, and implementing Internet of Things-based security systems. All these efforts aim to improve the overall library experience for users. The importance of simplifying academic libraries' operational procedures has emerged in recent years (Aslam et al., 2025). According to Aslam et al. (2025), an academic library is a specialized library associated with an educational institution that provides a multitude of information and services that support academic programs and research endeavors, making it an essential resource for researchers, faculty, and students. Academic libraries are adopting change management models more frequently, according to Cox (2023), in order to address the difficulties brought on by shifting user expectations, technological advancements, and changing educational environments. These change management models, according to Rožman et al. (2023), offer libraries organized frameworks to help them make the required adjustments. Therefore, by including library staff in the change process, Rahman et al. (2025) discovered that good change management increases staff engagement by creating a sense of ownership that lowers resistance, facilitates transitions, and boosts morale. Additionally, libraries can improve their services and make sure they better meet the needs of faculty and students by effectively implementing changes.

As academic libraries navigate the rapidly changing fields of education and research, they must actively embrace change to achieve their goals (Adams, 2024). To help these libraries deal with changes and transitions, change management models act as guiding frameworks. In addition to helping new processes or technology be successfully implemented, they also handle any resistance and encourage employee buy-in (Aslam et al. 2025). According to Adu et al. (2016), these models provide an organized approach that aids organizations in handling the challenges associated with implementing change. Frameworks for change management, according to Woodman (2024), focus on strategies for guiding an organization through a transition. Libraries in academic settings need carefully considered management strategies because of their critical role in education. These frameworks, according to Kossyva et al. (2024), provide a clear route from the current state to the intended future state, guaranteeing that changes occur with the least amount of disturbance.

Change management frameworks are crucial for academic libraries because of the constant changes in technology and education (Mızrak, 2024). Scholars have investigated a number of tactics to improve efficient change management in libraries. Ha and Thanh (2024) examined various frameworks for change management that are employed in school libraries. They discovered that models such as Kotter's Eight Steps were effective in enacting significant changes. In order to improve the efficacy of school libraries, they recommended modifying popular change management frameworks to suit their unique requirements. Similarly, Abukalusa and Oosthuizen (2025) recommended that library organizations employ frameworks that promote flexibility, adaptability, and ongoing education. The goal of this strategy is to make them more resilient to outside threats. Rosenbaum et al. (2018) emphasized how crucial it is for academic libraries to use change management frameworks. These frameworks assist libraries in adapting to the rapidly evolving needs of their patrons, technological advancements, and educational environments.

Personal Employee Development Plans (PEDPs) play a crucial role in handling change within academic libraries. They assist organizations to adapt to significant shifts and boost staff skills (Adeniyi et al., 2024). These plans help to create a workforce ready to face new challenges by focusing on personalized professional growth, which reduces resistance to change and sparks new ideas (Singh, 2023). Saul et al. (2024) point out that organized change management models have value because they offer ways to ease changes, cut down on disruptions, and keep stakeholders involved. As a result, blending tailored staff development plans with strategic change management boosts academic libraries' ability to adapt and prepares them to succeed amid ongoing shifts in information services and higher education.

Artificial Intelligence (AI) is revolutionizing employee development, according to recent studies. Madhumithaa et al. (2025) explore AI-powered innovations such as AI coaching tools, performance analytics, adaptive training modules, and smart learning management systems. According to their research, firms can identify skill gaps, design specialized training programs, and match workforce upskilling with strategic objectives by utilizing machine learning algorithms and predictive analytics. The study also emphasizes how AI can foster a culture of continuous professional development by increasing employee engagement through dynamic career planning tools and real-time feedback. Similarly, Khamis (2024) explored the usage of Degreed, an AI-enabled Learning Experience Platform (LXP) at a Swedish multinational corporation. Results indicated that this platform effectively integrates a variety of learning resources, offers personalized recommendations, and supports skill building aligned with what the organization needs. However, the study identifies some critical success factors for deployment, including making sure AI features align with employee learning preferences, preserving algorithmic decision-making transparency, and adhering to ethical data usage standards.

In order to increase staff productivity and unlock hidden potential, Maity (2019) emphasizes the necessity of personalized employee development plans. According to the findings, performance management procedures are always changing in modern organizations, and it is crucial to have well-designed, customized development plans that outline specific steps to improve performance. It suggests a shift from conventional systems of rewards and penalties to more comprehensive performance management strategies that emphasize employee growth. Patole (2024) empirically showed the real benefits of personalized employee development plans, such as better job performance, higher rates of retention, and greater workplace satisfaction. The authors argue that it is the ability to develop, rather than attract new talent, that will enable firms to survive in the increasingly fast-paced and diverse world of business. The study ends with a suggestion that organizations develop more global personalized employee development plans, which could lead to swift change of operations and technology. Gain.

1.1 Statement of the Problem

Academic libraries are known for enforcing change management due to a variety of resources and tools within them. Most academic libraries in Zimbabwe are struggling to put these strategies into practice. This could be caused by reduced training opportunities, resistance to change, and lack of fit between institutional objectives and librarians' skill sets, which are typically a big challenge. These challenges are mainly caused by a lack of personalized employee plans that can lead to poor change management initiatives and thereby lead to a decline in library services. The study aims to explore how employee development plans could assist librarians working in academic libraries in Zimbabwe with their change management models.

1.2 Research Questions

- How do personalized employee development plans influence the implementation of change management models in academic libraries in Zimbabwe?
- What are the perceived barriers to adopting personalized development plans among library staff?
- How can academic libraries in Zimbabwe effectively integrate PEDPs into their change management strategies?

2 Literature

2.1 Theoretical Framework

This study was inspired by the Lewin's Change Model. Kurt Lewin, a German-American psychologist in the 1940s, developed the Lewin's Change Management Model. The model remains relevant in today's evolving business world. It was characterized by three repetitive stages: unfreezing, changing, and refreezing (Ellis, 2023; Etareri, 2022). By creating awareness among the beneficiaries and aligning the stakeholders toward the change, the unfreezing stage reduces the resistance to the change; the changing stage is about implementing the new system using targeted interventions, such as resource allocation and training; and the refreezing stage consolidates these changes by embedding them in the culture of the organization (Rosdiana & Aslami, 2022). This model offers a structure for the embedding of PEPD plans as a strategic lever for change in the Zimbabwe academic libraries. Customized employee development plans can be of assistance in resisting unfreezing by creating buy-in through the alignment of librarians' career aspirations with the organizational goals (Koskei, 2024). As Jinga and Oumer Hassen (2025) note, during this process of refreezing, these changes become institutionalized by means of continuing feedback and reinforcement loops that range from peer mentoring and performance evaluations. Proactively providing professional development to staff differs from the changing phase, which does so through customized staff development plans and training, to possess the skills necessary to successfully transition to digital spaces or shifts in teaching practices. Beyond exercising operational inertia and addressing skill gaps, this approach enables libraries to remain nimble in the face of Zimbabwe's dynamic higher education environment, where technological advancement and resource constraints require agile, human-centered change management methods.

2.2 Personalized Employee Development Plans Influence the Implementation of Change Management Models

The success of change management models and approaches is also significantly affected by PEDPs. Madhumithaa et al. (2025) confirmed that personalized employee development plans had a statistically significant positive effect on employee-performance expectations, matching the change initiative. The outcome confirms the contribution of personalized development plans that are aligned with the purpose of the change initiative and that affect individual employees' positive intention to comply with change. Du Plessis and Mabunda's (2016) study focuses on the critical role of change management in academic libraries. They emphasized the importance of communication strategies in promoting change and mitigating resistance from academic library staff. Schroeder (2023) found that regular and ongoing assessments of development plans throughout their lifecycle would be required. This would enable organizations to respond to challenges as they arise and modify plans and change strategies where necessary to the change initiative. This change-responsive model will have a positive effect on the effectiveness of change management initiatives. Lejeune et al.'s (2016) study also confirmed the implementation of personalized employee development plans by managers in their organizations. Second, it confirmed the added value of personalized employee development plans for long-term learning goals and creating an efficient formal learning which was complemented by informal learning activities that were embedded and situated in the work itself.

In addition to the above, Widjaja (2023) concluded that by prioritizing personalized employee development plans, organizations can significantly improve the success of their change management initiatives, leading to smoother transitions, increased employee engagement, and improved organizational performance. The study also found the use of Awareness, Desire, Knowledge, Ability, and Reinforcement Model was useful as it focused on individual change as the foundation for organizational change, emphasizing the importance of awareness, desire, knowledge, ability, and reinforcement. Similarly, Hafez (2022) stated that personalized plans may reflect wider change objectives, thereby indicating where the employee development focus is to maximize an organization's success. The study found that Kotter's 8-Step Model offers a structured and strategic approach to change management and highlighted the importance of urgency, coalition, and communicating the vision. Likewise, Asniwati and Ramadhan (2025) concluded that personalized development plans created in response to personal needs and concerns helped reduce resistance to change because employees felt valued and supported. The study also provided evidence that the McKinsey 7-S Model emphasized the importance of aligning key elements of an organization (strategy, structure, systems, shared values, skills, style, and staff) in order to achieve successful change.

2.3 Barriers to Adopting Personalized Development Plans

Organizations, including academic libraries, face several barriers when trying to develop individualized employee training plans due to issues of operational, cultural, and systemic. The key problem is that neither managers nor staff understand the strategic importance of these plans, reducing participation. According to Brewster et al. (2016), this "wavering" apparently can be attributed to questions about the meaning of individualized development in the broader context of institution and (for that matter) the self. There is a dearth of training in the area of AI, because of a lack of new faculty hiring with expertise in computational sciences, as well as infrastructure in terms of computing resources and computer laboratories in the engineering schools, and structural impediments such as lack of autonomy of the educational institutions. Also, with decision-making centralized in the government, in a fixed curriculum format (Ahmed et al., 2019), inadequate upgradation of the technology in use, lack of response to the need to modernize teaching-learning processes (Barsha & Munshi, 2024). There are also ethical concerns around the digital divide with access as a result of a lack of internet or poor data plans (Barsha & Munshi, 2024). As well as allegations of easy lapses in data privacy, non-availability of digital infrastructure, lack of adequate numbers of skilled labor generated, and the high cost of putting into place AI-driven technologies (Barsha & Munshi, 2024).

Additionally, Kossyva et al. (2024) note that the resistance to change only aggravates these problems, but one reason workers refuse to change is that they often do not like changing because workers do not want to change their practices and they do not want to show that they need to learn something and fear the direction they want to take leadership, since it will have a better chance of a change, opening paths for administration to dismiss the workers. In addition, operational inefficiency arises due to no structured training for managers and employees in the development, deployment, and use of customized employee development plans. This has resulted in individuals being unable to form realistic objectives or identify relevant learning pathways (Koshanova et al., 2024). To be constantly self-improving takes quite a lot of motivation when you do not have the right environment. Limited access to resources, such as mentors, materials, and courses, has been identified as a barrier to the success of personal development initiatives (Maxwell et al., 2024). Based on Patole (2024), many individuals struggle to engage in their personalized employee development plan because of a lack of intrinsic motivation or unclear personal aspirations. The research also found that employees did not have enough options within the budget provided for classes or activities, and the limitations worked against their development.

2.4 Ways to Effectively Integrate Personalized Employee Development Plans into their Change Management Strategies

Developing a talent pool with the capacity to manage emergent needs requires consideration of personalized employee development plans as part of organizational change management. Thus, a tenuous link needs to be established between the integration of personal and organizational goals. For instance, Alabi et al. (2024) determined that when employee personal development goals are consistent with an organizational mission, employees are more motivated and engaged; thus, their intra-team efforts contribute to larger-scale change efforts. In addition, support from the top is essential as Lejeune et al. (2023) note that when management is transparent about the significance of personalized employee development plans, it legitimizes their position; thus, employees want to invest their customized efforts in the organizational mission.

Additionally, open lines of communication about potential changes reduce resistance and solidify buy-in, as suggested by Som and Chan (2024). For example, transparency about change implementation, trust through continued honest updates and planned feedback opportunities all suggest the appropriateness of personalized employee development plans for organizations perpetually in flux. Moreover, Schmitt and Pauknerova's (2026) study revealed that employees needed sufficient resources to cope with change as better coping techniques for distressing situations, better self-efficacy and self-awareness. In this regard, Holm et al. (2025) exposed the need for managerial support, training, and strategic implementation relative to the significance of the ongoing feedback system to maximize success. Thus, when rebranding performance management as a constant, organizations can energize employees, foster buy-in and promote sustained growth. This resonates with other ideologies that state that management should be the first to change behavior/learn something new (Mobolade & Akinade, 2021).

3 Methodology

A qualitative research method was adopted for data collection in this study, as it enabled the researchers to gather rich, in-depth information necessary to produce credible and meaningful results (Creswell et al., 2023). According to Dehalwar and Sharma (2023), qualitative methods are particularly effective when exploring complex social phenomena, such as the experiences and perspectives of individuals within specific contexts. In this case, the study focused on understanding the views and insights of senior librarians from state universities in Zimbabwe, a group with direct relevance and expertise concerning the study. A total of 20 library personnel were interviewed, representing all 10 state universities across the country. This coverage ensured that the data reflected a wide range of institutional experiences and perspectives, improving the study's credibility. Flick (2018) points out that looking at each person or object in a specific group can be the best way to gather reliable information, especially when the sample size is manageable, and each participant has valuable knowledge related to the research questions.

The participant selection process used a purposive sampling method, which involved intentionally choosing individuals based on their relevance and expertise in relation to personalized employee development plans and change management models within the university context. According to Creswell and Creswell (2022), purposive sampling involves handpicking participants to ensure that they can provide informed and insightful contributions, enhancing the quality of the qualitative data collected. The interviews were conducted using semi-structured guides, allowing for flexibility to explore emerging themes while keeping focus on the study's objectives. The data collected were then systematically analyzed using thematic analysis, a thorough method that involves identifying, coding, and organizing patterns and themes within the qualitative data (Creswell et al., 2023). This analytical method enabled a detailed understanding of the key issues, challenges, and opportunities related to personalized employee development plans and change management models in Zimbabwean state universities, ultimately adding to the study's overall credibility and depth.

4 Results and Discussion

4.1 What Is your Understanding of Personalized Employee Development Plans?

The data showed that most librarians understood what customized employee development plans were. One librarian said this about it: *"I think customized employee development plans are [probably] structures designed to help each employee achieve their professional goals and enhance their skills."* Customized employee development plans should be unique to each employee's strengths and weaknesses to enable a more focused professional development direction. The same type of themes appeared in participant responses in the way they understood a customized employee development plan. *"Customized employee development plans are a strategic document that outlines staff's individual training, learning, and career goals,"* an assistant librarian added. They ought to reflect the individual goals of each employee while aligned with the library's objectives. It is a way to ensure that everyone is developing in a way that supports both personal and organizational development. According to their comments, they all seem to understand customized employee development plans. This was validated by the following comments made by a Deputy Librarian: "*In my view, customized employee development plans play a critical role in developing a culture of continuous learning in the library. They provide staff with a structured approach to defining learning targets and tracking progress. By customizing these plans, we are better able to assist our employees in reaching their targets and, in turn, improve the resource efficiencies and quality of service of our library."* The majority of participants understood individualized employee development plans, which are crucial in academic libraries dealing with ever-changing technology and user expectations, according to the responses, which demonstrated a positive trend.

4.2 Can you Tell Is if you Have Used Personalized Employee Development Plans at your Libraries?

The use of personalized employee development plans in academic libraries was low in academic libraries. One Librarian said this: "*We have not yet implemented the use of personalized employee development plans at our library, but if implemented, they seem to be more useful and beneficial to the library as they can help our staff develop and do their duties in line with the intended objectives."* The use of personalized employee development plans was met with mixed feelings in some academic libraries. Another Sub Librarian revealed that: *"Yes, we have started to implement personalized employee development plans in our library. Initially, it was met with scepticism, but after a few workshops, many staff members saw the value. The plans help us identify specific training needs and career pathways. However, we still face challenges in terms of resources and time to dedicate to these plans."* In some libraries, they have shown that they never utilized personalized development plans at their libraries. A Librarian had this to say: *"Yes, we have not utilized personalized employee development plans in our library. But I know they are instrumental in helping staff identify their strengths and areas for improvement."* The results showed that some of the librarians have never used it, but others have shown that they were currently implementing it, which has revealed that despite its usefulness, the majority of libraries have not adopted the personalized employee development plans.

4.3 How Would you Describe the Current Change Management Models Used in your Library?

The participants showed that the libraries were using the top-down approach change management models, which were not all-inclusive. A Sub Librarian said this: *"Our library primarily follows a top-down change management model, where decisions are made by senior management and communicated to staff. While this approach helps in swift decision-making, it sometimes leaves little room for employee input, which can lead to resistance during significant changes."* Some of the library managers revealed that they used Lewin's Change Management model in their change management process. Another Librarian opined that: "*We largely rely on Lewin's Change Management Model, which includes unfreezing, changing, and refreezing. Other libraries have been using agile approaches to facilitate change processes."* One of the Deputy Librarians adds: *"Our library has been trying to adopt a more agile change management model. This means we focus on flexibility and responsiveness to change. We encourage team collaboration and regular feedback, allowing us to adapt our strategies based on staff input and emerging challenges. It's still a work in progress, but I believe it's effective in engaging our staff."* According to participant findings, different libraries use different change management models depending on the kind of change the library needs.

4.4 In What Ways Do you Believe Personalized Employee Development Plans Can Influence these Change Management Models?

The results from the interviews showed that personalized employee development plans could enhance change management models. An Assistant Librarian had this to say: *"I believe personalized employee development plans can significantly enhance change management models by fostering greater employee buy-in. When staff members see that their individual development is prioritized, they are more likely to embrace changes. PEDPs can also help identify specific training needs that align with the changes being implemented."* One of the librarians revealed that the personalized employee development plans could improve the skills of librarians in academic libraries. This was revealed by a Deputy Librarian who adds that: *"Personalized employee development plans can provide a structured way to address the skills gap that often arises during change initiatives. By focusing on individual learning objectives, we can tailor our change management strategies to better suit the needs of our staff, making the entire process smoother and more effective."* Some of the participants noted that when librarians are actively engaged in the need for change, they can easily overcome hurdles found in the change process. One of the Librarian opined that: *"I think personalized employee development plans can positively influence change management by promoting a culture of continuous improvement. When employees are actively engaged in their own development, they are more likely to contribute ideas and feedback during transitions. This involvement can lead to a more collaborative environment, which is essential for successful change management."* The findings revealed that the majority of participants believed that the use of personalized development plans can influence the change management models. This will allow a smooth transition in the change process.

4.5 What Do you Perceive as the Main Barriers to Adopting Personalized Development Plans among Library Staff?

Participants indicated that there was a lack of awareness regarding the use of personalized employee plans among librarians. They felt that they were not beneficial on their part. A Sub Librarian said this: *"I think one of the main barriers is a lack of awareness about what personalized development plans entail. Many staff members do not fully understand their benefits or how to create them. Additionally, there is often a limited amount of time for staff to focus on their own development amidst their daily responsibilities."* A key barrier to the use of personalized employee development plans was a lack of support from the library management and critical stakeholders within the university. Another Assistant Librarian opined that: *"From my perspective, a key barrier is the insufficient support from management. If leadership does not prioritize or invest in personalized development, it can send a message that it is not important. Moreover, workload pressures often take precedence over personal development, making it difficult for staff to engage fully with their plans."* Some of the library staff mentioned the issue of underfunding to support the adoption of personalized employee development plans. A librarian adds that: *"A significant barrier is the resource constraint we face. Implementing personalized development plans requires time, training, and sometimes funding for courses or workshops. Without adequate resources, it's challenging to make these plans effective. There is also some resistance to change; not everyone is comfortable with the idea of a structured development plan."*

4.6 What Best Practices Would you Recommend for Effectively Integrating PEDPs into Change Management Strategies?

The findings from the participants revealed the need to use a clear-cut communication structure that gives room for library staff to understand the reasons for the adoption of personalized employee development plans. An Assistant Librarian had this to say: *"I would recommend starting with clear communication about the purpose and benefits of PEDPs. Providing training sessions for staff on how to create and implement these plans can help. It is also important to ensure that everyone understands how their individual goals align with the library's overall change objectives."* The top managers saw the idea of buy-in and involvement among library staff as critical. A Sub Librarian adds that: *"Engaging staff in the development of their own personalized employee development plans is crucial. I think involving them in the process fosters ownership and commitment. Regular check-ins to discuss progress and adjust the plans as needed can also keep everyone aligned and motivated during changes."* Supporting staff was seen as key through the use of appropriate change management models. One of the Deputy Librarian opined that: *"Creating a supportive environment is key. Leadership should actively promote the use of personalized employee development plans, providing resources and time for staff to focus on their development. I also suggest integrating feedback mechanisms to gather insights from staff about their experiences with personalized employee development plans during change initiatives."* Linking personalized employee development plans with objectives to facilitate change was seen to produce positive results. A Librarian also said this: *"To effectively integrate personalized employee development plans into change management, I recommend linking individual development goals to specific change initiatives. This way, employees can see the direct relevance of their plans. Additionally, celebrating small successes and milestones can boost morale and reinforce the value of the development process." The results from participants showed a mixed approach in the use of strategies that could help integrate personalized employee development plans with change management models."* The use of clear communication with library staff, regular check-in of progress with staff, a supportive environment where staff could engage their managers, and linking the personalized employee development plans with their task objectives was seen as the best strategy to be employed by library managers in academic libraries.

The findings on the personalized employee development plans showed that the library staff have a broad understanding of the importance of these plans in promoting individual growth and ultimately enhancing organizational effectiveness. One of the participants concurred that tailored frameworks allowed employees to achieve their career aspirations while aligning with the library's goals. This dual focus helps to create a learning culture; structured plans help staff set clear learning objectives and track their progress. Madhumithaa et al. (2025) supported this, stating that individualized employee development plans align change-related organizational objectives with employees' personal goals. Where this correspondence exists, organizations may expect that employees will take on a more positive role in the change processes.

However, the implementation of personalized employee development plans faces several barriers within library settings. It was also revealed that structural limitations further complicate adoption because of resource constraints, infrastructural gaps, a shortage of skilled personnel, and the prohibitive costs (Barsha & Munshi, 2024). Structural limitations create different facets toward adoption because it creates different facets in the way it has been led. A major typical theme among the participants is the extent to which these plans are unknown and that a lack of understanding hinders adoption. Insufficient support from management and resource constraints (time and funding) further complicate efforts to integrate personalized development into existing frameworks. Staff find it challenging to prioritize their professional development due to resistance to change and current workload pressures, underscoring the need for strategic leadership support. According to Kossyva et al. (2024), this makes these problems worse because workers are frequently afraid to alter established procedures, fear that skill gaps will be discovered, or doubt the intentions of leaders due to concerns about their job security.

Participants suggested a number of best practices for successfully incorporating customized employee development plans into change management tactics. Training sessions that give staff members the skills they need to design and carry out their own plans are crucial, as is clear communication regarding the goals and advantages of customized employee development plans. Bah et al. (2024) stress that leadership commitment is equally important because it not only validates the value of individualized employee development plans but also encourages employees to match their developmental endeavors with organizational priorities. Regular check-ins can help sustain motivation and alignment with organizational changes, while involving employees in the development process promotes ownership. Furthermore, tying personal objectives to particular change projects guarantees pertinence and promotes employee involvement. This aligns with existing research findings by Alabi et al. (2024), who indicated that, when employees' personal development objectives are directly linked to the organization's mission, they exhibit increased motivation and engagement, ensuring that their contributions complement larger change initiatives. By highlighting the importance of individualized development in the library setting, milestone celebrations can further boost morale. All things considered, achieving the maximum potential of customized employee development plans requires a cooperative and encouraging atmosphere.

5 Implication of the Study

Personalized employee development plans in academic libraries support the Sustainable Development Goals (SDGs) by connecting staff growth with the goals of quality education (SDG 4), decent work and economic growth (SDG 8), and reduced inequalities (SDG 10). These plans improve change management by preparing staff for evolving library roles. They ensure ongoing professional growth that helps deliver sustainable services and strengthens institutional resilience. This alignment enhances the effectiveness of change management models, as staff development directly supports overarching institutional objectives, leading to improved service delivery (Schroeder, 2023). This connection supports policy development that focuses on professional development and readiness for future challenges. It positions academic libraries as important contributors to the SDGs through their development of human capital.

6 Limitations and Areas of Further Research

The study may not have taken into consideration the particular dynamics and difficulties found in other kinds of libraries or institutions because it focused exclusively on academic libraries in Zimbabwe. The results might not apply to special libraries, public libraries, or libraries in other cultural or economic contexts. There is a need for further studies regarding the effects of customized employee development plans in academic libraries. Studies should focus on longitudinal studies.

7 Conclusion

The study found that PEDPs are key to making academic libraries in Zimbabwe adapt change management models. Customized employee development plans help librarians to actively face change by linking their personal growth goals with those of the organization. This gives them a sense of ownership and makes them less likely to fight the new plans. Customized training and development give the librarians the skills they need to change with the new ways of doing things and new tools, increasing their ability to meet the needs of those who use their services. In this way, academic libraries are better able to face change and keep their power and meaning in the learning world. The results also showed how important it was for leaders to be committed and to have a strong corporate culture of ongoing learning and professional growth. Library managers can create an environment where staff can grow and try new ways of working by actively pushing and adopting customized employee development plans into change management strategies. This daily approach not only helps libraries to be ready for change but also helps to grow a better, more flexible team. This final way shows how the study shows how academic libraries in Zimbabwe could greatly benefit from customized staff development plans, making way for better service and better results in a quickly changing world of learning.

Recommendations

The study made the following recommendations:

- The study suggested that academic libraries include individualized staff development plans in their change management initiatives to improve librarians' capacity for change adaptation.
- The study recommended that academic library management should provide robust training for librarians to develop competencies needed for effective change management.
- The study recommended that academic librarians should include all stakeholders for the effectiveness of personalized employee development plans in supporting change management models.
- The study also suggested that academic libraries make use of digital tools and platforms to support the execution of customized employee development plans.

References

Abukalusa, K., & Oosthuizen, R. (2025). An adaptive organisational leadership framework through systems thinking. *International Journal of Organizational Analysis, 33*(8), 2673–2703.

Adams, R. J. (2024). *Information technology & libraries: A future for academic libraries*. Routledge.

Adams, M. J., Finch, M., & Gamble, B. (2023). More than a game: Scenario planning, imagination, and the public Library's future. *Public Library Quarterly, 42*(6), 665–687.

Adeniyi, I. S., Al Hamad, N. M., Adewusi, O. E., Unachukwu, C. C., Osawaru, B., Onyebuchi, C. N., & David, I. O. (2024). Organizational culture and leadership development: A human resources review of trends and best practices. *Magna Scientia Advanced Research and Reviews, 10*(1), 243–255.

Adu, A. O., Abbas, J., & Evonlah, N. (2016). The nexus of quality management and construction Organisations' performance in urban Ghana: The moderating effect of information technology integration. *International Journal of Innovative Science and Research Technology, 9*(4), 1972–1988.

Ahmed, F., Qin, Y. J., & Martínez, L. (2019). Sustainable change management through employee readiness: Decision support system adoption in technology-intensive British E-businesses. *Sustainability, 11*(11), 2998. https://doi.org/10.3390/su11112998

Akuffo-Aduamah, C. C. N. (2025). The impact of compensation on employees performance in organizations. *International Journal of Multidisciplinary Studies and Innovative Research, 13*(1), 57–69.

Alabi, O. A., Ajayi, F. A., Udeh, C. A., & Efunniyi, C. P. (2024). Data-driven employee engagement: A pathway to superior customer service. *World Journal of Advanced Research and Reviews, 23*(3), 923–933.

Alam, A. F., Subaveerapandiyan, A., Mvula, D., & Tiwary, N. (2024). AI literacy and Zambian librarians: A study of perceptions and applications. *Open Information Science, 8*(1), 20220166.

Aslam, M. (2022). Adapting to change in academic libraries. *Global Knowledge, Memory and Communication, 71*(8–9), 672–685.

Aslam, S., Jabeen, M., & Jabeen, M. (2025). *Fostering innovation in smart library services: An in-depth exploration of empowered capabilities and dynamic strategies among university librarians in Pakistan.* Information Development, 02666669241305379.

Asniwati, A., & Ramadhan, A. N. J. (2025). Employee management during times of organizational change: Strategies for overcoming resistance and work stress. *Economics and Digital Business Review, 6*(1), 997–1014.

Babatope, V. O. (2025). Mentoring strategies for enhancing work-life balance in academia: Supporting Well-being and productivity. In *Strategic mentoring in higher education: Best practices for academic success* (pp. 59–75). Springer Nature Switzerland.

Bah, M. O. P., Sun, Z., Hange, U., & Edjoukou, A. J. R. (2024). Effectiveness of organizational change through employee involvement: Evidence from telecommunications and refinery companies. *Sustainability, 16*(6), 2524.

Bandura A, A (2024). Bandura, A. (2023). Cultivate self-efficacy for personal and organizational effectiveness. *Principles of organizational behavior: The handbook of evidence-based management* (3rd ed., pp. 113–135). https://doi.org/10.1002/9781394320769.ch6.

Barsha, S., & Munshi, S. A. (2024). Implementing artificial intelligence in library services: A review of current prospects and challenges of developing countries. *Library Hi Tech News, 41*(1), 7–10.

Brewster, C., Gooderham, P. N., & Mayrhofer, W. (2016). Human resource management: The promise, the performance, the consequences. *Journal of Organizational Effectiveness: People and Performance, 3*(2), 181–190.

Burke, W. W. (2023). *Organization change: Theory and practice.* Sage.

Chychun, V., Grechanyk, O., Khliebnikova, T., Temchenko, O., & Kravchenko, H. (2023). Change management models and methods: Implementing innovations, ensuring sustainability and engaging staff. *Economic Affairs.*

Cox, J. (2023). The position and prospects of academic libraries: Strengths and opportunities. *New Review of Academic Librarianship, 29*(4), 367–393.

Creswell, J., Poth, C. N., & Rawlins, P. (2023). Mapping design trends and evolving directions using the sage handbook of mixed methods research design. *The Sage handbook of mixed methods research design*, 527–537.

Crowley, M. C. (2025). *The power of employee Well-being: Move beyond engagement to build flourishing teams.* Berrett-Koehler Publishers.

Dehalwar, K., & Sharma, S. N. (2023). *Fundamentals of research writing and uses of research methodologies.* Edupedia Publications Pvt Ltd.

Du Plessis, T., & Mabunda, T. T. (2016). Change management in an academic library in the knowledge economy. *South African Journal of Libraries and Information Science, 82*(1), 53–61.

Ellikkal, A., & Rajamohan, S. (2025). AI-enabled personalized learning: Empowering management students for improving engagement and academic performance. *Vilakshan-XIMB Journal of Management, 22*(1), 28–44.

Ellis, P. (2023). Managing change—The Lewin model 3: refreeze. *Wounds UK, 19*(4), 69–71.

Errida, A., & Lotfi, B. (2021). The determinants of organizational change management success: Literature review and case study. *International Journal of Engineering Business Management, 13*, 1–15. https://doi.org/10.1177/18479790211016273

Etareri, L. (2022). An analysis framework of change management. *Medicon Engineering Themes, 3*(1), 30–38.

Flick, U. (2018). *An introduction to qualitative research* (6th ed.). https://uk.sagepub.com/en-gb/eur/an-introduction-to-qualitative-research/book261109

Fredberg, T., & Pregmark, J. E. (2022). Organizational transformation: Handling the double-edged sword of urgency. *Long Range Planning, 55*(2), 102091.

George, T. J., Maneethai, D., Atwater, L. E., & Madera, J. M. (2025). When remote work works better: How fit and autonomy shape employee outcomes. *Organizational Dynamics*, 101175.

Ghedabna, L., Ghedabna, R., Imtiaz, Q., Faheem, M. A., Alkhayyat, A., & Hosen, M. S. (2024). Artificial intelligence in human resource management: Revolutionizing recruitment, performance, and employee development. *Nanotechnology Perceptions, 20*(S10), 52–68.

Ha, D. V., & Thanh, N. C. (2024). Applying the Kotter change management model in high school education: Some initial results in Hanoi. *Vietnam. European Journal of Education Studies, 11*(8), 81–96.

Hafez, M. (2022). *Strategic change leadership: Aligning team efforts with organizational goals for sustainable change.* Available at SSRN, 4802280.

Holm, C. G., Kringelum, L., & Anand, A. (2025). Creating effective strategy implementation: A systematic review of managerial and organizational levers. *Review of Managerial Science, 20*(2), 673–705.

Jinga, A. A., & Oumer Hassen, J. (2025). Extent of leadership roles in change management process in public universities of Ethiopia: A mixed-methods study. *Cogent Education, 12*(1), 2453284.

John, J. (2010). Relevance of employee development plans in employee performance management. *SSRN Electronic Journal.* https://doi.org/10.2139/ssrn.1711647

Kalandarovna, A. G., & Qizi, A. M. A. (2023). Development and increase of competitiveness of the organization. *ASEAN Journal of Educational Research and Technology, 2*(3), 265–274.

Khamis, R. (2024). *AI-powered learning experience platforms: Investigating personalized learning in the workplace.* University of Gothenburg. https://hdl.handle.net/2077/83632

Koshanova, D., Zakirova, A., & Tynyshkali, A. (2024). *Development of a personalised learning pathway on the basis of a competency-based approach. Ensaio: Avaliação e Políticas Públicas em Educação, 32*(124).

Koskei, J. (2024). Analysis of change management facilitation for library system migration in academic libraries in Kenya: A case of Moi university, Eldoret. *Eastern African Journal of Humanities and Social Sciences, 3*(1), 161–170.

Kossyva, D., Theriou, G., Aggelidis, V., & Sarigiannidis, L. (2024). Retaining talent in knowledge-intensive services: Enhancing employee engagement through human resource, knowledge and change management. *Journal of Knowledge Management, 28*(2), 409–439.

Lejeune, C., Beausaert, S., & Raemdonck, I. (2023). Effective personal development plans contextualized: The role of the autonomy-supportive people manager in sustaining employees' self-directed learning. *Human Resource Development Quarterly, 34*(4), 389–412.

Lejeune, C., Mercuri, D., Beausaert, S., & Raemdonck, I. (2016). Personal development plans supporting employee learning and perceived performance: The moderating role of self-directedness. *Human Resource Development International, 19*(4), 307–328. https://doi.org/10.1080/13678868.2016.1203639

Madhumithaa, N., Sharma, A., Adabala, S. K., Siddiqui, S., & Kothinti, R. R. (2025). Leveraging AI for personalized employee development: A new era in human resource management. *Advances in Consumer Research, 2*, 134–141.

Maity, S. (2019). Identifying opportunities for artificial intelligence in the evolution of training and development practices. *Journal of Management Development, 38*(8), 651–663.

Maxwell, B., Káplár-Kodácsy, K., Hobson, A. J., & Hotham, E. (2024). Effective mentor training, education and development: A qualitative multilevel meta-synthesis. *International Journal of Mentoring and Coaching in Education, 13*(4), 422–440.

Mizrak, F. (2023). Enhancing organizational competitiveness through entrepreneurial education, human resources management, and innovation strategies: A Vikor and Fuzzy AHP approach. *International Journal of Social and Humanities Sciences Research (JSHSR), 10*(100), 2525–2545

Mızrak, F. (2024). Effective change management strategies: Exploring dynamic models for organizational transformation. In *Perspectives on artificial intelligence in times of turbulence: Theoretical background to applications* (pp. 135–162). IGI Global Scientific Publishing.

Mızrak, F., & Caylan, S. (2023). Human resource training and its effect on the integration and success of organizational change. In *Managing successful and ethical organizational change* (pp. 120–143). IGI Global Scientific Publishing.

Mobolade, G. O., & Akinade, M. E. (2021). Team building and teamwork in organizations: Implications to managers and employees in work places. *International Journal of Management, Social Sciences, Peace and Conflict Studies, 11*(4), 1.

Oyedokun, T. T. (2025). Navigating the dynamics of present-day academic libraries: An in-depth analysis of strategies, challenges, and emerging trends. *IFLA Journal, 51*(2), 470–489.

Patole, S. R. (2024). The role of ethical and trustworthy AI teammates in enhancing team performance: A systematic literature review. *Performance Improvement Quarterly, 37*(2), 53–66.

Phillips, J., & Klein, J. D. (2023). Change management: From theory to practice. *TechTrends, 67*(1), 189–197.

Platt, M. (2020). *The leader's brain: Enhance your leadership, build stronger teams, make better decisions, and inspire greater innovation with neuroscience.* University of Pennsylvania Press.

Rachmad, Y. E. (2022). *Change management theory.* Tilburg Textiel Boek Internationale Uitgeverij, Speciale Editie 2022.

Rahman, M. H., Ghazali, A. B. M., & Sawal, M. Z. H. B. M. (2025). Exploring organizational factors of resistance to technology adoption in university libraries in Bangladesh. *Information Development*, 02666669251325447.

Rosdiana, N., & Aslami, N. (2022). The main models of change management in Kurt Lewin's thinking. *Jurnal Akuntansi, Manajemen Dan Bisnis Digital, 1*(2), 251–256.

Rosenbaum, D., & More, E. (2022). Towards a strategic change management framework for the nonprofit sector: The roll-out of Australia's national disability insurance scheme (NDIS). In *Effective implementation of transformation strategies: How to navigate the strategy and change interface successfully* (pp. 415–450). Singapore.

Rožman, M., Tominc, P., & Štrukelj, T. (2023). Competitiveness through development of strategic talent management and agile management ecosystems. *Global Journal of Flexible Systems Management, 24*(3), 373–393.

Saul, T.F., Binza, M.S., & Moeti, K. (2023). An alignment of personal development plans with training and development of employees at SASSA. *SA Journal of Human Resource Management/ SA Tydskrif vir Menslikehulpbronbestuur, 21*(0), a2161. https://doi.org/10.4102/sajhrm.v21i0.2161

Saul, T. F., Binza, M., & Moeti, K. (2024). A model for optimizing the functional relationship between personal development plans and training and development plans: A case of the south African social security agency. *IAHRW International Journal of Social Sciences Review, 12*(4), 461–467.

Schmitt, J., & Pauknerova, D. (2026). Building adaptive leaders: Rethinking leadership development in the age of uncertainty. *European Journal of Training and Development, 50*(3–4), 293–312.

Schroeder, C. (2023). *Coming in from the margins: Faculty development's emerging organizational development role in institutional change.* Taylor & Francis.

Singh, J. (2023). Change Management in the Digital era: Overcoming resistance and driving innovation. *Journal of e-Science Letters, 4*(3), 7–11. https://doi.org/10.51470/eSL.2023.4.3.07

Som, R., & Chan, R. (2024). Examining employee buy-in for organizational change: A structural equation model. *Studies in Business and Economics, 19*(2), 250–262.

Sripada, C. (2024). *Shaping the future of work: Build flexible work options and unleash the human Capital of Bharat.* Penguin Random House India Private Limited.

Swanson, R. A. (2022). *Foundations of human resource development.* Berrett-Koehler Publishers.

Tait, E., Martzoukou, K., & Reid, P. (2016). Libraries for the future: The role of IT utilities in the transformation of academic libraries. *Palgrave Communications, 2*(1), 16070.

Uka, A., & Prendi, A. (2021). Motivation as an indicator of performance and productivity from the perspective of employees. *Management & Marketing, 16*(3), 268–285.
Vezi, P. T., Rajlal, A., Chitamba, A., & Muchowe, R. (2024). Examining the nexus between staff development practices and employee performance at a selected municipality in KZN. *International Journal of Business Ecosystem & Strategy, 2687-2293, 6*(6), 91–101.
Weinper, K., & Pietrzyk-Leonowicz, S. (2022). The academic library during the process of change. Change process, disruption, and evaluation. *Library Philosophy and Practice*, 7292. https://digitalcommons.unl.edu/libphilprac/7292
Widjaja, P. (2023). Effective employee management in times of organizational transformation. *Advances in Human Resource Management Research, 1*(3), 137–149.
Woodman, R. W. (2024). Creativity and organizational change: Linking ideas and extending theory. In *Handbook of organizational creativity* (pp. 283–300). Psychology Press.

Janeth Matingwena is a Lecturer in Human Resource Management at Manicaland University of Applied Science in Zimbabwe. She holds a DPhil in Human Resource Management from the Chinhoyi University of Technology, along with postgraduate and undergraduate qualifications in Human Resource Management and a Diploma in Adult and Continuing Education. Her research focuses on Human Resource Development, Organizational Behavior, Emotional Intelligence, Employee Resourcing, and Decent Work, with particular emphasis on higher education and workforce performance. She presented her work at international conferences on topics including doctoral supervision, supervisor support, and human resource development. She had also reviewed several papers and published in peer-reviewed journals. Dr Matingwena has participated in several international conferences and recently completed a certified research supervision training program from Lindi University, Sweden, enhancing her engagement with global academic scholarship.

Admore Mashokoh is an Assistant Librarian in the Scholarly Communication Section at Great Zimbabwe University Library. He holds a Bachelor of Science Honors Degree in Library and Information Science. Mr Mashokoh is an active researcher and has authored several peer-reviewed journal articles and book chapters in Library and Information Science, published with reputable academic publishers. His scholarship and professional interests span Scholarly Communication, Academic Research Support, and Human Resource Development in librarianship, with a particular focus on capacity building, staff training, and professional growth within academic libraries.

Exploration of Artificial Intelligence As a Catalyst for Human Resource Management Effectiveness

Abimbola Foluso Ojapinwa, Rasheed Olawale Azeez, and Foluso Ilesanmi Jayeoba

Abstract The role of Artificial Intelligence (AI) as a driving force behind efficient human resource management was examined in this chapter. Effective performance management and the recruitment and selection process of getting the best-fit candidate are paramount toward organizational performance. Failure in this regard may result in loss of competitive edge in the market. Adopting artificial intelligence in performance management and the recruitment and selection process is not only more efficient and precise but also fairer and with less bias, as they are free of human feelings. This research investigated the relationship between predictive performance analytics and performance management. It also examined the effect of the robotics automation process on the recruitment and selection process. Descriptive research design was adopted in this study using 148 workers of BeaconGate Limited as the study population. The Statistical Package for Social Sciences [SPSS], version 24, was used to design and assess two study hypotheses using linear regression analysis and the Pearson Product-Moment Correlation technique. Research findings confirmed a correlation between the utilization of predictive performance analytics and performance management and likewise an influence of robotics process automation on the recruitment and selection process. This study offered insightful information about how to integrate AI technology into an organization's human resource management strategy for effectiveness. This study, therefore, recommended amongst others that human resource practitioners need to encourage integration of AI software (robotics automation, performance analytics) in their practices to mitigate the drawbacks of conventional human resource management (recruitment and selection process, performance management) systems, as this will aid effective human resource management in various organizations.

A. F. Ojapinwa (✉) · R. O. Azeez · F. I. Jayeoba
Department of Industrial Relations and Human Resource Management, Lagos State University, Ojo, Nigeria
e-mail: abimbola.ojapinwa@lasu.edu.ng

R. O. Azeez
e-mail: rasheed.azeez@lasu.edu.ng

F. I. Jayeoba
e-mail: foluso.jayeoba@lasu.edu.ng

M. Dzingirai and S. O. Atiku (eds.), *Strategic Human Resource Analytics and Intelligent Data-Driven Systems*, https://doi.org/10.1007/978-981-92-2623-8_6

Keywords HRM · AI tools · Robotics automation · Recruitment and selection · Performance management · Performance anagement analytics

1 Introduction

Employers are under more pressure than ever due to high unemployment and the need to fill new roles effectively, efficiently, and promptly with the best-fit candidates despite a skilled labor scarcity. When a job position is posted, the human resource department receives a large volume of applications from candidates, and it might be difficult to dedicate the necessary time and effort to review each applicant's resume (Raed et al., 2025). According to Sha (2024), resume screening and traditional human resource management (manual resume screening, face-to-face interviews, subjective tests, regular reviews, and casual notes) have long been labor-intensive, time-consuming, and extremely tiresome. They are also vulnerable to human mistakes and subjective prejudice. Fallucchi et al. (2020) observed that subjective assessments, biases, and a lack of data are some of the fundamental drawbacks of traditional performance evaluation techniques.

Semwal et al. (2024) noted that traditional performance appraisal mechanisms are often anchored in subjective perception and inherently vulnerable to cognitive bias and evaluative prejudice. Majority of HR-related tasks and processes, like hiring and choosing employees, managing performance, training and developing staff, have traditionally been difficult and onerous (Elenwo, 2025). These constraints inhibit accurate projections and informed decision-making. Artificial intelligence integration is changing the paradigm of HR processes, which historically depended largely on manual labor, subjective judgment, and time-consuming procedures. AI offers data-driven insights, automates repetitive tasks, and allows HR professionals to contribute more strategically (Md. Masudul et al., 2025). To increase productivity and effectiveness, lessen recruiting biases, and improve decision-making, AI-driven recruitment systems make use of machine learning, robotics automation, and both predictive analytics and natural language processing (Venkateshwaran et al., 2025).

Vijai and Mariyappan (2023) documented that robotics automation in human resource management lowers human error, boosts process effectiveness, and protects data integrity. This makes it possible for HR professionals to use analytics to glean insightful information from HR data. Tanisha et al. (2025) posited that predictive analytics has emerged as a crucial instrument that enhances the effectiveness and decision-making process of Human Resources (HR) management by employing data, statistical techniques, and machine learning to identify patterns and make educated guesses about future events. Businesses will be more equitable and effective in terms of recruitment and selection and performance assessments by eliminating human biases and relying on data-driven insights (Lilia et al., 2024). The integration of AI and predictive analytics will enable HR to move from reactive to proactive performance management, enhancing decision quality and organizational agility. This will also reduce the possibility of partiality and ensure that employees are assessed according to their real work and contribution level. Despite all the benefits of AI integration toward achieving effective human resource management, Vishesh (2021) opined that there remains a possibility that HR may not have fully investigated AI (robotics automation, predictive analytics) to fully understand its positive impact on effective human resource management. Likewise, Steven and Na (2022) stated that the potential influence of Human Resource (HR) analytics on organizational performance and effectiveness is yet uncertain, despite the field's expansion and use. Comprehending the underlying reasons, mechanisms, and timing through which HR analytics enhances organizational performance, which can be gauged by effective human resource management, is therefore essential. Based on the foregoing, this study explored how artificial intelligence (robotics automation and performance analytics) serves as a catalyst for effective human resource management.

1.1 Statement of the Problem

Human Resource Management (HRM) has experienced tremendous reforms recently in Nigeria and all over the world based on technological innovations. Nevertheless, human resource personnel/managers are still experiencing some setbacks, leading to inefficiency and ineffectiveness in the organization. Besides, this results in subjective prejudice, non-transparency, unaccountability, biases, unfair recruitment and selection processes, and poor performance management. Akwei and Nwachukwu (2023) believed that situational constraints like inadequate digital uptake, meddling by the politicians, lack of regulatory structures for strategic HR transformation, and infrastructure deficiencies exacerbate these difficulties, especially in Nigeria. According to Saini and Khurana (2023), the swift integration of AI and technological advancements has revealed gaps in HR professionals' digital literacy which limits the strategic use of data-driven workforce planning. Rogovsky and Cooke (2021) noted that although businesses are adopting digital technology and artificial intelligence with enthusiasm in many HRM spheres, their comprehension of how these advancements impact employees is frequently not prioritized and falls below expectations. This is a wakeup call on organizations to capitalize and organize constant digital mastery training initiatives to improve human resource professionals' capability to use AI (robotics automation, performance analytics, and so on) effectively. Building digital HR infrastructure through collaboration between the private and public sectors may enhance data management and performance-driven cultures, especially in Nigeria.

Besides, conventional managerial operations in human resource processes must give way to Human Resource approach that values people first, uses technology to make processes easier, and relies on proven evidence to guide fair and effective decisions. To increase organizational legitimacy, policy changes that enforce fair recruitment and selection, accountability, and openness in human resource procedures are crucial.

1.2 Hypotheses of the Study

$H_{1:}$ There is a significant effect of robotics process automation on recruitment and selection practices.

$H_{2:}$ There is a significant association between predictive performance analytics and performance management.

2 Conceptual Literature

2.1 Artificial Intelligence

Artificial intelligence originated in the twentieth century. It is the application of technology to replicate human cognitive abilities while also adding the capacity to foresee and account for possible issues (Rathore, 2023). According to Raed et al. (2025), artificial intelligence, as a byproduct of industrial technology, has impacted every part of our existence and led to a contempt for life in social work, humanitarian efforts, and overall society, in addition to science. It is the study of concepts that are introduced to machines so that they can react to stimuli in a manner similar to that of humans. These concepts include the ability to think, judge, and have the willpower to analyze, criticize, and choose among several viewpoints within themselves. Based on UNICEF (2021a, 2021b), in Elenwo (2025), AI is a machine-based system that can directly or indirectly provide a virtual setting, forecasts, make suggestions, or a set of human-defined goals. The time spent on discrete and labor-intensive jobs is greatly decreased by the AI technologies, which carry out hiring procedures that previously needed human judgment (Asajile et al., 2024).

2.1.1 Predictive Performance Analytics

The word predictive analytics means the usage of statistical methods and modeling to predict future outcomes and performance (Poddar & Chattopadhyay, 2021). Vishesh (2021) discovered its autonomous or semi-autonomous data analysis or knowledge employing cutting-edge technology, typically used outside of traditional Business Intelligence (BI), and processes to reveal deeper insights, generate forecasts, or recommendations. Tanisha et al. (2025) established that predictive analytics uses machine learning, statistical techniques, and historical data to identify patterns and make educated guesses about the future. Its primary objective is to make educated, data-driven decisions for businesses before they occur by speculating about potential outcomes or human behavior. By analyzing current and historical data patterns, predictive analytics helps businesses and investors identify potential future trends and make informed decisions. Gupta (2023) recognized that using data and sophisticated algorithms to find patterns and extract insightful information, machine learning and business analytics approaches can improve the precision of performance forecasts.

2.1.2 Robotic Process Automation

This is the term used to describe the automation of repeatable and rules-driven operations in corporate processes using software robots, or "bots"(Sharma, 2025). By imitating how human users interact with computer systems, these bots may complete activities like data entry, data extraction, form filling, and so on. The repetitive and routine work currently performed by humans may be replaced by a "robot" through Robotic Automation of Processes (RPA). Naveen (2023) opined that by automating repetitive operations, RPA has become a potent tool to tackle some issues like business landscape, organizations going through constant pressure to enhance service delivery, lower expenses, and increase operational effectiveness.

2.1.3 Human Resource Management

Employees are an organization's most significant asset as they contribute to its objectives both individually and collectively. An intentional and coordinated strategy for managing these resources is Human Resource Management (HRM). It is the art of finding, training, and keeping skilled employees to accomplish company objectives as efficiently as possible. Managing employees and maximizing performance are the responsibilities of human resource management and are crucial organizational functions (Elenwo, 2025). It covers the protocols, rules, and methods for recruiting, educating, hiring, evaluating, keeping, and bringing on the proper mix of employees with the right skills to meet the goals of the company. Human resource management deals with raising an organization's productivity through procedures, technologies, and concepts.

2.1.4 Performance Management

Performance management entails setting standards and goals for the workplace, monitoring performance against them, actively managing all performance levels, and optimizing learning and growth (CIPMN, 2018). The goal of performance management is to quantify and enhance employee performance. Lekshmi and Prakash (2024) established that performance management is the essence of HR which helps in enhancing individual and team performance to improve organizational performance. Considering the latest shift in company goals, Tatineni and Allam (2024), cited in Beatrice and Joanes (2025), claim that the necessity to assess employee performance has led to the need to restructure the workforce. Therefore, in order to help employees understand what they need to do and why, performance management ought to be incorporated into the business transformation program.

2.1.5 Recruitment and Selection Process

The process of captivating, locating, and evaluating possible applicants for open positions in a company is known as recruitment. It includes several tasks, like posting jobs, screening candidates, conducting interviews, and determining which candidates are qualified, experienced, and a good fit for the role. Nassar et al. (2025) pointed out that the four primary steps of recruitment and selection are candidate screening, candidate matching, candidate sourcing, and assessment of the applicant. Accordingly, selection is the process of choosing the best candidate from a group of applicants for a certain job opening. It usually entails a thorough assessment of applicants' abilities, credentials, experience, and cultural fit using techniques like assessments, reference checks, background checks, and interviews. It is a process where a firm must identify those skill sets required by employees to be successful. Onyeche and Deedam (2023) recognized that recruitment is an essential process in the prosperity of any organization. It deals with the determination of the proposed number of positions to be filled in an organization. Likewise, Onwuchekwa, cited in Onyeche and Deedam (2023), posited that the process by which an organization chooses from a pool of candidates' applications or applicants who best meet the qualifications for the available position while taking the current situation of the market into consideration is known as selection. It is about concluding who to choose.

2.1.6 Adoption of AI in Human Resource Management

The adoption of Human Resource Management (HRM) processes within emerging economies has progressed from basic administrative tasks to technologically powered, strategic approaches targeted at boosting efficiency, competitiveness, and effectiveness. To promote effective hiring, employee engagement, and data-driven decision-making, digital HR tools like Human Resource Information Systems (HRIS), Artificial Intelligence (AI), and analytics-based performance systems have been increasingly integrated in recent years (Singh et al., 2023).

Artificial Intelligence (AI) solutions in HRM provide information-driven choices in workforce analytics, performance management, and hiring (Davenport et al., 2020). The adoption of AI in HR parallels wider trends in digital transformation. The prospect for HR transformation via automation, analytics, and decision support is demonstrated by AI's worldwide dissemination. Though its spread differs widely owing to contextual variables including infrastructure, regulation, and institutional preparation (OECD, 2023). For instance, the use of AI in Zimbabwe is still growing, mostly experimental, and only found in a small number of metropolitan firms. According to Newsday Zimbabwe (2024), businesses in the banking, telecom, and retail industries are now adopting AI-enabled HR tools for recruitment, payroll, and workforce analytics. AI has been more prominently integrated into HR and company processes in Southern African nations like South Africa, Namibia, and Botswana than in others. Nonetheless, SADC Secretariat (2023) documented that the digital gap continues to be a barrier throughout the Southern African Development Community (SADC) as AI testing is limited outside big urban regions because many firms lack cloud infrastructure/data governance capabilities.

PwC (2023) explained that investment, supporting regulations, and robust digital infrastructure are driving the adoption of AI across North America and Europe. For instance, 41% of United States major firms reported utilizing AI for HR-related tasks in the year 2023, and Asia-Pacific nations, including China, Japan, and Singapore, have included AI into their national digital policies. For example, China's "New Generation AI Development Plan" places a strong emphasis on HR analytics and labor automation (Zeng et al., 2022).

Accordingly, the implementation of HRM techniques in Nigeria is indicative of a hybrid system that is marked by both enduring structural inefficiencies and new digital advancements. Even while top companies have started using digital performance management tools, online training, and artificial intelligence hiring, many continue to function inside antiquated bureaucratic structures that impede strategic HR change (Akwei & Nwachukwu, 2023; Onifade & Adewale, 2024).

AI is a crucial component of HR innovation worldwide, as Zimbabwe and several countries are making progress, but they are limited by institutional and infrastructure issues, digital illiteracy among HR professionals, inadequate infrastructure, cultural reluctance to change, inadequate financing for digital efforts, and a lack of regulations to govern the ethical use of technology (Khan, 2024; Suresh & Shanmugapriya, 2024). A unified initiative is needed in Nigeria and other developing countries to close the disparity in the adoption of technology and HRM effectiveness. This includes public–private investment in stable technological infrastructure, upskilling of human resource managers in analytics and AI governance, and enforced policies that establish ethical and transparent utilization of HR technologies (Onifade & Adewale, 2024; Tan & Lim, 2023). Once utilized, these actions have the potential to change fragmented adoption into HRM practices that are responsible, inclusive, and performance-driven.

2.2 *Empirical Review*

2.2.1 Artificial Intelligence and Human Resource Management

The use of artificial intelligence in human resource management to enhance organizational performance was examined by Singh et al. (2023). An interdisciplinary, thorough review of over fifty current studies was carried out by the experts. Although AI has been found to offer several advantages to HRM, there are very few concerns with its application in the workplace. A new era of HR (version 5.0) was ushered in with the usage of AI technology in HRM, which is novel in the modern world. The analysis reached a conclusion that it will help the economy in the long run and produce a significant number of new jobs. Elenwo (2025) determined and investigated the obstacles preventing Nigerian higher education institutions from successfully integrating artificial intelligence into twenty-first-century human resource management. These consist of a lack of technical know-how, inadequate facilities, prohibitive expenses for implementation, worries about data privacy and security, staff resistance to change, restricted funding options, a lackluster policy and regulatory structure, a reliance on manual procedures, a lack of AI-powered HR tools, and limited awareness and comprehension of the advantages of AI. The study concluded that the effective implementation of artificial intelligence in the twenty-first century is severely hampered by these limitations, resulting in a lack of competitiveness, inefficiency, and ineffectiveness. Training and development programs, infrastructure improvements, cost–benefit evaluation, safeguarding information policies, change management, funding research, development of policies, awareness creation, automation of procedures, and vendor collaboration are some of the strategies the study recommended to address these limitations. Higher education institutions in Nigeria will have more leverage if these limitations are addressed.

Asajile et al. (2024) evaluated how Artificial Intelligence (AI) affected the recruiting process for a subset of Non-Governmental Organizations in Dar es Salaam's Kinondoni municipality. In particular, the study evaluated how AI affects communication, applicant engagement, and selection. Primary and published secondary data from the internet were utilized to obtain data for this study, which used a descriptive case study methodology. According to the findings, advancements and discoveries have led to computers, robots, and other artificial intelligence possessing human-like intelligence, which is typified by cognitive capacities, learning, flexibility, and decision-making abilities. The study's conclusions also showed that AI-based HRM systems may help to achieve several goals, such as cost reduction, hiring delays, and efficient resource usage. Since the AI system runs online, the study suggests that the Ministry of Communication lower data costs.

Lilia et al. (2024) assessed the usage of AI in HRM by modernizing recruitment, performance, and employee development. The scholars identified that artificial intelligence is used in identifying resumes, evaluating candidates, and, at times, interviewing candidates via the chatbot. They concluded that the usage of AI made it possible to screen the right candidate to be hired, avoid biases that come with hiring based on physical appearance, track employees' performance impartially by giving real-time feedback, and finally, estimate trends in workers' performance for meaningful decision-making.

2.2.2 Predictive Performance Analytics and Performance Management

Performance management has been transformed by combining technology and data analytics, using AI and machine learning to examine enormous datasets and find trends in worker performance (Obi, 2024). Raed et al. (2025) investigated how employee experience can be enhanced, thereby increasing the effectiveness of human resource management using artificial intelligence via performance analysis of businesses. The study evaluated how AI affected companies that used technology to improve employee contentment and maximize HRM effectiveness. A cross-sectional descriptive research design was used in this study, and data from the respondents was gathered all at once. One hundred and eighty foreign human resources specialists who worked for companies with entities in the US, the United Arab Emirates, and Jordan were among the participants. The scholars discovered the emerging benefits of AI for HRM and emphasized the need for safety and confidentiality of data, to control any modifications that AI applications may cause for staff members and look for reasonably priced AI solutions tailored to the HR department's requirements.

Based on the Technology Acceptance Model (TAM) and Systems Theory theoretical frameworks, Beatrice and Joanes (2025) provided an in-depth assessment of empirical research on artificial intelligence effectiveness over organizational performance methods of management between 2020 and 2024. This analysis emphasizes the ways in which AI technologies are incorporated into performance management systems to boost productivity, lessen prejudices, and increase worker engagement. Several empirical studies examining AI applications in performance management were analyzed using a systematic literature review methodology. It was found that AI makes it easier to create individualized development plans, decrease biases in evaluations, enable insight based on data, and support continuous feedback loops. The scholars concluded that predictive analytics are essential for predicting worker performance and spotting future leaders.

Halawi et al. (2024) presented a conceptual article on how human resource analytics affects the organizational performance of both business and non-profit enterprises as well as the elements of employee performance and behavior. In contrast to non-profit firms, which showed no correlation between HR analytics and employee performance and behavior, profit organizations demonstrated a favorable association between HR analytics and employee performance and behavior.

Obi (2024) illustrated how important data analytics and technology are to the transformation of performance management systems. The study brought attention to the transition from conventional approaches to more contemporary, flexible, and data-driven procedures while highlighting how new technologies have the potential to further transform the industry. The study emphasized how crucial it is to implement these innovations to improve organizational competitiveness in the digital era, employee happiness, and management effectiveness. The study concluded that, despite their drawbacks, old performance management techniques served as a basis for contemporary systems.

Yuxin et al. (2024) suggested an approach for production management that uses RFID data to analyze worker performance. This strategy uses RFID data to analyze workers' performance. The original source material was long and complicated, including parts that were troublesome and not finished. The data for the specified criterion is chosen using computer programming, and a Python program was then used to analyze and condense the data into five criteria. The first research question was then addressed by analyzing and visualizing the resultant data. Finally, data comparable to group 1 was used to test the method. In addition to fulfilling the study goal, this experiment put forth two novel ideas. First off, the analytic techniques produced useful insights from the RFID data, making it possible to evaluate staff performance. Quantified KPIs are obtained utilizing a variety of methods and programming languages; all outcomes are computed using data and formulae without the need for human review. This strategy works better for businesses that need more talent and is simpler to replicate. Secondly, while the test group did well, the data application was not an isolated instance. For validation, a proportional third of the employees and the associated facts are chosen at random from the group. It turns out that the performance in the test group and the separated group 1 have a parallel trend and were readily separated from the other four groups using the same procedure. The scholars recommended that, based on the result, the model may be applied once more if comparable data from the same factory becomes available.

Steven and Na (2022) identified the processes via which improved performance happened. The scholars significantly tackled the reasons, methods, and timing of how HR analytics enhance organizational performance. Organizational performance, HR analytics, HR technology, and Evidence-Based Management (EBM) were all assessed using the chain mediation model via structural equation modeling, and data were gathered from 155 Irish firms. The study's results provided credence to the suggested chain model, showing how HR analytics are made feasible by having access to HR technology, which facilitates EBM and enhances organizational performance.

The integration of HR analytics with performance management was investigated by Lekshmi and Prakash (2024). To understand the publishing pattern, prevailing topics, and upcoming research trends in this field, the study uses bibliometric analysis. The focus of the study was on examining the advantages and consequences of using analytics in performance management. VOSviewer is used to analyze bibliometric data from the Scopus database. The researchers claim that contemporary research was focused on the use of predictive analytics in training, retention, attrition, and employee performance. It was discovered that there isn't much research that analyzes the influence of analytics in performance management functions experimentally.

Lekshmi and Prakash (2024) explored HR Analytics integration in performance management. The study relies on bibliometric analysis to comprehend the publication pattern, dominating themes, and forthcoming research trends in this area. The study's concentration was on exploring the benefits and effects of analytics on the application of performance management. Bibliometric data from the Scopus database was subjected to analysis using VOSviewer. According to the scholars, recent studies are keen on the application of predictive analytics in employee performance, attrition, training, and retention. It was found that studies that empirically measure the impact of analytics in performance management function are limited in number. In order to close this gap and address the need for studies that develop and suggest change methods for the successful application of analytics, the study suggested further research. The study came to the conclusion that the use of descriptive, predictive, and prescriptive analytics is necessary for the successful execution of a strategic function involving performance management.

2.2.3 Robotics Process Automation and Recruitment and Selection Process

Venkateshwaran et al. (2025) assessed a study that implements an integrated methodology to examine the effect of Automated Reasoning Systems (ARS) in employee resourcing within the Industrial Technology industry. The study used both quantitative and qualitative methods of data collection to guarantee a thorough comprehension of automated recruiting procedures. This qualitative and quantitative design focuses on the advantages, difficulties, and ethical issues of AI's transformation of hiring and selecting employees. Statistics were generated via structured questionnaires disseminated to IT staff, hiring managers, and HR specialists. Also, secondary data was used. For example, moderated interviews with AI recruiting experts and managers of human resources provided fundamental knowledge of how AI works in the hiring process. This study assessed how to comprehend the trends and difficulties of AI-based hiring, consulting current studies, sector analysis, contextual review, and corporate human resources guidelines. Judgmental sampling technique was adopted by picking human resource specialists and workers in the information technology sector with knowledge in AI-powered hiring. Dimensions involved 150 respondents, with 15 HR professionals interviewed. Correlation as a research technique for data analysis was analyzed utilizing the Statistical Package for Social Sciences (SPSS), and interview responses were analyzed via thematic analysis. It was found that by increasing productivity, decreasing prejudice, and raising applicant happiness, the IT business will undergo significant change due to Artificial Intelligence (AI) in terms of hiring and selection procedures.

The hiring process is streamlined by AI-powered technologies such as chatbots, prediction analytics plus monitoring systems for applicants, which make it quicker and more efficient in finding the top applicants. The evaluation revealed that higher employee satisfaction is correlated with AI deployment, and more seasoned HR professionals are more likely to see AI technologies favorably since they can better comprehend their benefits and handle their drawbacks. The study concluded that, though AI provides considerable advantages in hiring, industries must implement an equitable strategy that guarantees moral behavior, and the reduction of errors by people with ongoing advancements in algorithms in ensuring that AI tools are transparent and fair for optimization while curtailing potential risks.

Sha (2024) examined how artificial intelligence is used in hiring and choosing candidates. According to the scholar, AI has a significant impact on competency evaluation, interviews, and resume screening during the recruitment and selection process. Businesses may increase recruiting accuracy, efficiency, and screen qualified candidates more quickly by implementing AI technology. Additionally, the study established that with technologies like speech and picture recognition, which have greater promise in competence evaluation, AI can assist businesses in better evaluation of individuals' skills and potential during the interview process. The study recommended that to better integrate artificial intelligence into the enterprise's recruitment and selection process and to support efficient management, enterprise managers should keep researching and summarizing the technology's significant role in these processes.

Asajile et al. (2024) evaluated how artificial intelligence (AI) affected the recruiting process for a subset of NGOs in Dar es Salaam's Kinondoni municipality. In particular, the study evaluated how AI affects communication, applicant engagement, and selection. Primary and published secondary data from the internet were utilized to obtain data for this study, which used a descriptive case study methodology. According to the findings, advancements and discoveries have led to computers, robots, and other artificial intelligence possessing human-like intelligence, which is typified by cognitive capacities, learning, flexibility, and decision-making abilities. The study's conclusions also showed that AI-based HRM systems may help achieve a number of goals, such as cost reduction, hiring delays, and efficient resource usage. Given that an AI system runs online, the study suggests that the Ministry of Communication lower data costs.

Nassar et al. (2025) treated how AI-powered technologies are currently being developed in connection with the hiring and selection process. Numerous AI-based applicant sourcing solutions, including Fetcher, hireEZ, Textio, Entelo, and XOR, were found to exist, according to the review. The employer may choose from a pool of applicants created by this AI-powered technology without the need for human participation. The techniques identified the best candidates for open positions. Additionally, the AI-based applicant screening systems that mainly use chatbot technology to help and guide candidates during the employment process include Mya system, Wade and Wendy, AllyO, Humanly Lisa, Debra, Brazen, Olivia, Ideal, and Eightfold.

Additionally, these AI-powered tools facilitate candidate matching and ensure that the company hires the best candidate by matching the prospective employee's specifications and job requirements with respect to the minimum qualifications, the job location, the skills required, the amount of time spent working, and the salary requirements. According to the results, AI-powered systems for evaluating candidates may also analyze speech patterns, body language, and candidate interactions. It was suggested, therefore, that more study be done on the subjective criteria's accuracy and quality.

To investigate the benefits plus disadvantages of utilizing AI in hiring and selection, Ore and Sposato (2022) conducted research that looked at recruitment specialists' perspectives. An exploratory inquiry was conducted using a qualitative methodology. Ten seasoned recruiters from global firms participated in in-person interviews. The results demonstrated how AI automates routine task performance. Recruiters are cautious and distrustful of AI technology because of the hazards involved in its usage in recruitment and selection. If implemented properly, AI can improve hiring procedures. Although there is still pessimism due to worries about widespread unemployment brought on by technology, the respondents think that their jobs would survive because recruiters should always be human.

The purpose of Asajile et al.'s (2024) study was to evaluate how Artificial Intelligence (AI) affected the recruiting process for a subset of NGOs in Dar es Salaam's Kinondoni municipality. In particular, the study evaluated how AI affects communication, applicant engagement, and selection. Primary and published secondary data from the internet were utilized to obtain data for this study, which used a descriptive case study methodology. According to the findings, advancements and discoveries have led to computers, robots, and other artificial intelligence possessing human-like intelligence, which is typified by cognitive capacities, learning, flexibility, and decision-making abilities. The study's conclusions also showed that AI-based HRM systems may help achieve a number of goals, such as cost reduction, hiring delays, and efficient resource usage. Due to the fact that the AI system runs online, the study suggests that the Ministry of Communication lowers data costs.

2.2.4 Theoretical Framework

Technology–Organization–Environment (TOE) Framework

The 1990 study of Tornatzky and Fleischer served as the foundation for the Technology–Organization–Environment (TOE) concept. This information systems model was developed to show how a number of variables, including the characteristics of the technology, the organizational setting in which it is utilized, and the external environment in which the organization functions, impact the uptake and application of new technologies. Based on the swift development and wide application of AI, several institutions have initiated the use of AI in HRM (Sha, 2024). With some specific applications of AI technological tools such as robotics automation and predictive performance analytics in HRM in terms of recruitment and selection and performance management, there have been positive outcomes such as efficiency, productivity, and promotion of customer satisfaction in several organizations. The robotics automation process permits HR professionals to concentrate on tactical initiatives that add value to the organization (Naveen, 2023). Its assistance in the provision of the most suitable candidates for job requirements and in decision-making support cannot be overemphasized during the interview process.

Additionally, AI can simulate real work scenarios and assess candidates' abilities and adaptability with the help of technologies with mixed realities and a simulated environment. Venkateshwaran et al. (2025) documented that AI-driven recruitment systems leverage machine learning, predictive analytics, and Natural Language Processing (NLP) to boost effectiveness, lessen hiring biases, and refine decision-making. Sharma (2025) believed that Robotics Process Automation (RPA) has progressed from automating routine admin duties to assisting with increasingly intricate operations, including hiring, onboarding, employee engagement, and retention tactics. Lekshmi and Prakash (2024) posit that organizations using traditional approaches in performance management are confronted with obstacles such as autonomy, discontinuity, and inefficient data utilization. The scholars advised organizations to integrate analytics for an evidence-based approach to decision-making, and likewise, a better way of HR data usage. Furthermore, encouraging an open and adaptive organizational culture will boost experimentation and innovation which is crucial toward the sustenance of a positive culture throughout digital transformation

Table 1 Population size

Department	Number of staffs	Male	Female
Manufacturing	127	121	6
Marketing and sales	35	8	27
Operations	50	39	11
Admin, HR, and others	23	18	5
Total	235	186	49

Source Researcher's (2025)

3 Methodology

This study was hinged on the descriptive research design with a population that included 235 workers of BeaconGate Limited, a consulting company in Lagos, Nigeria. Table 1 shows the population size for this study.

Stratified random sampling technique was used to split the organization into strata, based on departments, and each department represented a stratum. The departments are Operations, Marketing & Sales, Human Resources, Manufacturing, and others. The questionnaire served as a means of data collection and sample selection was based on the willingness to represent the population efficiently. A sample size of 148 respondents, representing 36.55% of the population, was chosen using the Yaro Yamani (1967) formula. Both descriptive and inferential statistics were adopted for this study and data was analyzed using Pearson's correlation and version 25 of the Statistical Package for Social Sciences (SPSS) to perform linear regression.

4 Results

4.1 Hypotheses Testing

Hypothesis one,

> H_1: *There is a significant effect of robotics process automation on recruitment and selection practices.*

Table 2 shows the results of linear regression analysis on the influence of robotics process automation on recruitment and selection practices.

Table 2 Regression analysis results

Model	R	R square	Adjusted R square	Std. Error of the estimate
1	0.701a	0.491	0.487	0.414973

a. Predictors: (constant), robotics process automation

Table 3: Coefficients for Robotics Process Automation and Recruitment and Selection Practices Model.

The regression coefficient results indicated that robotics process automation positively and significantly influenced recruitment and selection practices ($\beta = 0.826$, $p = 0.000$). This revealed that R-squared is 0.491 which presupposes that robotics process automation explains 49.1% of the variation in recruitment and selection practices. This indicated that a unit increase in robotics process automation would result in a 0.826 increase in recruitment and selection practices. The alternative hypothesis, H1, was accepted while the null hypothesis, Ho, was rejected.

H_2: There is a significant association between predictive performance analytics and performance management.

Table 4 presents the result of correlation between Predictive Performance Analytics and Performance Management

Table 4 revealed the result of a test on the relationship between predictive performance analytics and performance management which is moderately positive and statistically significant ($r = 0.660$, $p < 0.05$). Consequently, the alternative hypothesis was accepted and the null hypothesis was rejected because there is a noteworthy connection between predictive performance analytics and performance management. This showed that a rise in predictive performance analytics usage would lead to an increase in performance management.

Table 3 Regression coefficients

Model	Unstandardized coefficients	Standardized coefficients	t	Sig.	
B		Std. error		Beta	
1	(constant)	0.515	0.260	1.980	0.050
Robotics process automation	0.826	0.071	0.701	11.621	0.000

a. Dependent variable: Recruitment and selection practices

Table 4 Pearson correlations

Correlations

		Predictive performance analytics	Performance management
Predictive performance analytics Artificial intelligence	Pearson correlation	1	0.660^{**}
	Sig. (2-tailed)	.	0.000
	N	100	100
Human resource practices	Pearson correlation	0.660^{**}	1
	Sig. (2-tailed)	0.000	.
	N	100	100

5 Discussion

This study focused on the effects of robotic process automation on recruitment and selection practices, as well as the linkages between performance analytics and performance management. Considering the results, there is an influence of robotic process automation on recruitment and selection practices. It is believed that robotics automation reduces hiring biases and enhances efficient decision-making and competence evaluation. Using robotics automation in competence evaluation can not only make the outcomes more objective and accurate, but it can also save time and money. HR managers may do their assessment jobs more quickly and give workers immediate feedback on their findings for their career development with the help of automated and smart assessment technologies (Sha, 2024). According to Venkateshwaran et al. (2025), subjective decision-making and interviews frequently result in biases, inefficiencies, and delays in the hiring of talent. AI-powered hiring platforms provide a data-driven strategy that improves the hiring process's effectiveness, precision, and equity. Applicant Tracking Systems (ATS), learning through machines techniques, predictive analytics, and Natural Language Processing (NLP) are primary mechanisms of robotics automation in hiring workers.

Also, the results revealed a healthy correlation between performance analytics and performance management. Tanisha et al. (2025) posited that the way modern firms operate has been totally transformed using predictive analytics to assess employee performance. Using concepts drawn from data, it improves the accuracy, equity, and efficiency of crucial HR activities. Businesses move past outdated methods of making snap judgments about individuals based on feelings due to predictive analytics. This outcome is consistent with Vishesh's (2021) findings that recommended the use of predictive analytics as its application in HRM is very substantial. Furthermore, the Technology–Organization–Environment (TOE) framework (1990) supports the findings of this study as it elucidates the endorsement and application of technological innovations in organizations.

6 Contributions of the Study

This study emphasizes how crucial the need to combine individual knowledge with technology to guarantee efficient HRM procedures. To HR practitioners and business experts, these findings offer helpful insights into the strategic implementation of AI technology to enhance effective HRM. This study provided insightful information academically by adding to the expanding corpus of research on the impact, relationship between AI and HRM likewise, by illuminating certain domains in which AI technologies have the potential to revolutionize HRM. Time-to-hire and the likelihood of errors in hiring are greatly decreased, and the applicant experience is improved with quicker answers and more transparent updates when robotics automation is employed to simplify procedures such as applicant communication, appointment scheduling, and assessment of resumes. HR practitioners need to encourage the integration of performance analytics in their practices to mitigate the drawbacks of the conventional performance management system.

7 Limitations and Areas for Further Research

This research concentrated solely on the influence of artificial intelligence (robotics automation, performance analytics) on human resource management (recruitment and selection process, performance management). It is also limited because of the setting, sector, and company being studied. The study's conclusions might not be as broadly applicable as they could be. When using AI technologies in HRM, different sectors, corporate cultures, and demographics may provide varied outcomes. Additionally, there were only 148 participants in the sample, which might not be entirely representative of the general population. Research could be carried out on the influence of AI on other HRM operations such as compensation management, talent management, and so on. Tackling these topics can further deepen our knowledge and support the continuous development of effective HRM in the world of artificial intelligence.

8 Conclusion

This study focused on the linkages between AI and HRM and likewise, the impact of AI on effective HRM. The study contributed to previous knowledge by emphasizing the importance of incorporating AI in ensuring effective human resource management. AI has become a disruptive tool that is changing how human resource management is carried out in the contemporary workplace. AI can increase productivity, accuracy, and support in making effective decisions while lowering costs across a variety of processes, including hiring, selection, and performance management. Human resources management has been made easy because of the development of robotic process automation and performance analytics, which have increased productivity and decreased costs. This study, therefore, concludes that there is a positive relationship and a positive impact of AI on HRM.

References

Akwei, C., & Nwachukwu, C. (2023). An exploration of contextual factors affecting the nexus of competitive strategy and human resource management practices in Nigeria emerging economy context. *The International Journal of Human Resource Management, 34*(16), 3079–3122.

Asajile, M., Silverius, K., & Crispin, M. (2024). Influence of artificial intelligence on selection stage of recruitment in Tanzania: A case of selected NGOs in Kinondoni municipality. *International Journal of Research and Innovation in Social Science (IJRISS), 8*(12), 3406–3416.

Beatrice, C., & Joanes, K. (2025). Performance management and artificial intelligence (AI): Enhancing personalized development with continuous feedback and data-driven decisions. *International Journal of Computer Science and Information Technology Research, 13*(1), 13–22.

Davenport, T. H., Guha, A., Grewal, D., & Bressgot, T. (2020). How artificial intelligence will change the future of marketing. *Journal of the Academy of Marketing Science, 48*(1), 24–42.

Elenwo, P. M. (2025). Constraints of artificial intelligence integration in the 21st century human resource management in higher institutions in Nigeria. *International Journal of Educational Management, Rivers State University., 1*(1), 278–287.

Fallucchi, F., Coladangelo, M., Giuliano, R., & William De Luca, E. (2020). Predicting employee attrition using machine learning techniques. *Computers, 9*(4), 86.

Gupta, P. (2023). HR analytics: An effective tool for performance management and business development. In *HR analytics in an era of rapid automation* (pp. 112–126).

Halawi, A., Rasheed, R., & Al Belushi, B. (2024). The effect of human resource analytics on employee performance. *Revista de Gestão Social e Ambiental, 18*(5), 1–12.

Khan, A. M. (2024). The effects of brain drain on economic growth and human capital: Implications for developing countries. *Journal of Global & Social Issues, 3*(1), 1–14.

Lekshmi, C. R., & Prakash, P. (2024). Analytics integration in performance management: A bibliometric analysis. *International Journal of Multidisciplinary Research and Analysis, IJMRA, 7*(5), 2045–2055.

Lilia, Rania, Qanita, Muhammad, Ahmad, Mohammed & Shahadat (2024). Artificial intelligence in human resource management: Revolutionizing recruitment, performance, and employee development. *Nanotechnology Perceptions, 20*(10), 52–68.

Md. Masudul, H. B., Kripa, N. D., Palash, S., Pankaj, K. S., Halimuzzaman, M., & Md Tanjil, B. (2025). Exploring the role of artificial intelligence in transforming HR practices. *International Journal of Business Management and Economic Review, 8*(1), 2581–4664.

Nassar, S., Jaradat, N., Ajouz, M., & Alzeaideen, K. (2025). The impact of multiple intelligences on recruitment and selection of employees. In A. Hamdan, A. Harraf, & P. Arora (Eds.), *From machine learning to artificial intelligence: The modern machine intelligence approach for financial and economic inclusion* (pp. 601–612).

Naveen, E. V. (2023). Automating HR processes with robotic process automation (RPA*). Journal of Engineering and Applied Sciences Technology, 5*(1), 1–5.

Obi, R. I. (2024). Leveraging technology and data analytics in performance management-an exploratory study on its evolution: The when, now, and hereafter. In *ARPHA conference abstracts* (Vol. 7, p. e129581). Pensoft Publishers.

OECD. (2023). *Artificial intelligence in business and finance: Global adoption trends*. OECD Publishing.

Onifade, T. A., & Adewale, M. O. (2024). Artificial intelligence and HR analytics: A framework for HRM transformation in Nigeria. *African Journal of Business and Management Research, 5*(2), 56–69.

Onyeche, C., & Deedam, D. G. (2023). Recruitment and selection processes in public service organizations: Rivers State experience. *Global Academic Journal of Humanities and Social Sciences, 5*(2), 123–133.

Ore, O., & Sposato, M. (2022). Opportunities and risks of artificial intelligence in recruitment and selection. *International Journal of Organizational Analysis, 30*(6), 1771–1782.

Poddar, A., & Chattopadhyay, S. (2021). A proposed theory for predicting employee performance using machine learning. In *Interdisciplinary research in technology and management* (pp. 511–516). CRC Press.

PwC. (2023). *AI predictions 2023: Global insights*. PwC Research Institute.

Raed, H. W., Fida, Z., Leila, R., & Emad, A. (2025). Utilizing artificial intelligence to enhance employee experience and improve human resource management efficiency: A performance analysis of companies. *Journal of Ecohumanism, 4*(1), 5355–5379.

Rathore, S. P. S. (2023). The impact of AI on recruitment and selection processes: Analysing the role of AI in automating and enhancing recruitment and selection procedures. *International Journal for Global Academic & Scientific Research, 2*(2), 78–93.

Rogovsky, N., & Cooke, F. L. (2021). Human resource management and technological change: Emerging challenges and future directions. *Journal of Management Development, 40*(6), 541–558.

Saini, R., & Khurana, R. (2023). Reducing bias in recruitment and selection through digital transformation. *Asian Journal of Management and Human Resource, 10*(1), 33–47.

Secretariat, S. A. D. C. (2023). *Digital transformation and innovation report 2023*. SADC Publications.

Semwal, R., Tripathi, N., Rana, A., Pandey, U. K., Parihar, S., & Bairwa, M. K. (2024, September). AI-Driven insights: Enhancing personality type prediction with advanced machine learning algorithms. In *2024 7th International Conference on Contemporary Computing and Informatics (IC3I)* (Vol. 7, pp. 815–822). IEEE.

Sha, R. N. (2024). Application of artificial intelligence in recruitment and selection. *Academic Journal of Science and Technology, 9*(2), 56–58.

Sharma, V. (2025). The evolution of robotic process automation in human resources: From recruitment to retention. *International Journal of Multidisciplinary Research and Growth Evaluation, 5*(5), 1053–1058.

Singh, S., Thakur, P., & Singh, S. (2023). How does the use of AI in HRM contribute to improved business performance? A systematic review. In *Managing Technology Integration for Human Resources in Industry 5.0*, 131–139.

Steven, M. C., & Na, F. (2022). Bridging the gap: Why, how and when HR analytics can impact organizational performance of trinity business school, Trinity College Dublin. *The Unive Management Decision, 60*(13), 25–47.

Suresh, R., & Shanmugapriya, S. (2024). The combination of artificial intelligence (AI) and automation is altering the human resource management environment. *International Journal of Research in Human Resource Management, 6*(2), 419–423.

Tan, L., & Lim, H. (2023). Technology adoptionin HRM: Barrires and opportunities in emerging economies. *Asia—Pacific Journal of Management Research, 40*(1), 88–104.

Tanisha, S. L., Ratnaprabha, R. B., Shital, S. B., Vaidehi, S. K., & Ravindra, S. A. (2025). The impact of predictive analytics on employee performance evaluation and succession planning in modern organizations. *Journal of Information Systems Engineering and Management, 10*(1).

Tatineni, S., & Allam, K. (2024). AI-driven continuous feedback mechanisms in devops for proactive performance optimization and user experience enhancement in software development. *Journal of AI in Healthcare and Medicine, 4*(1), 114–151.

Tornatzky L.G., Fleischer M., Chakrabarti A.K., 1990, *The processes of technological innovation Issues in organization and management series,* Lexington Books. Retrieved June 10, 2013, from http://www.amazon.com/Processes-Technological-Innovation-Organization/Management/dp/0669203483

Unicef (2021a). Global Forum on AI for Children. *Virtual conference | Addressing pressing issues at the intersection of children's rights, digital technology and AI systems.*

Unicef (2021b). Policy guidance on AI for children. *Ministry of foreign affairs Finland*. https://www.unicef.org/globalinsight/media/2356/file/UNICEF-insight-policy

Venkateshwaran, G., Rajesh Kumar, N., Luyang, & Vishnupriya, S. D. (2025). Artificial intelligence in HR: Transforming recruitment and selection in IT industry. *Journal of Information Systems Engineering and Management, 10*(17), 2468–4376.
Vijai, C., & Mariyappan, M. (2023). Robotic process automation (RPA) in human resource functions. *Advances in Management, 16*(3), 30–37.
Vishesh, G. (2021). The significance of predictive analysis in predicting the performance of members in the organizations. *International Journal of Recent Technology and Engineering (IJRTE), 9*(5), 51–56.
Yuxin, W., Yishu, Y., & Ray, Y. Z. (2024). Data analytics for working performance analysis in production management. *Manufacturing Letters, 41*, 73–80.
Zeng, Y., Lu, E., & Huangfu, C. (2022). China's new generation AI development plan: Challenges and opportunities. *AI & Socliety, 37*(2), 733–748.
NewDay Zimbabwe. (2024, February 18). *AI reshaping the future of HR management in Zimbabwe.* https://www.newsday.co.Zw

Abimbola Foluso Ojapinwa is a researcher specializing in Entrepreneurship Development, Human Resource Soft-Skills, and Human Resource Management.

Rasheed Olawale Azeez is a scholar in Human Resource Management and Occupational Well-Being, whose research explores the human dimensions of urban transport systems in African cities. His work lies at the intersection of human resource management, employee well-being, and inclusive urban mobility, with a particular focus on transport workers.

Foluso Ilesanmi Jayeoba is a Lecturer in the Department of Industrial Relations and Human Resource Management, Faculty of Management Sciences, Lagos State University, Ojo, Lagos, Nigeria. He specializes in Human Resource Management, Entrepreneurial Development Studies, and Organizational Psychology.

Ethics in Human Resource Analytics: Perspectives from Public and Private Organizations in Ghana

Richard Afedzie and Eugene Okyere Kwakye

Abstract This chapter examines the ethical ramifications of Human Resource Analytics (HRA) within Ghana's public and private sectors, focusing on justice, privacy, transparency, and ethical governance. There is no gainsaying that employers across the globe are interested in making decisions based on data. Similarly, there has not been enough research on the ethical aspects of HRA in less developed countries. Utilizing stakeholder theory, ethical decision-making theory, and socio-technical systems theory, this study employed a qualitative exploratory design within an interpretivist framework. Forty (40) human resource professionals from twenty (20) Ghanaian businesses were interviewed using semi-structured protocols. We used Braun and Clarke's (2006) six-step thematic analysis to make sure the data was reliable and complete. Five major problems came up. These included algorithmic bias and fairness, data privacy and surveillance, employee autonomy and permission, profiling and behavioral trust, and organizational culture and ethical governance. The results showed that in Ghanaian businesses, the lack of digital infrastructure, low levels of data literacy, and weak ethical frameworks made discrimination and privacy violations more likely. Cultural hierarchy and regulatory gaps often make it harder to be open and responsible when using analytics. However, some companies that are ahead of the curve, especially in banking and telecommunications, have started to include fairness audits, ethical review processes, and compliance with Ghana's Data Protection Act (2012, Act 843). The study contributes to the evolving African discourse on ethical analytics by illustrating that ethical governance, trust, and cultural alignment are essential for responsible data practices. It finds that in Ghana's changing HR analytics landscape, capacity training, ethical leadership, and policy reinforcement are all important for keeping innovation and integrity in balance.

Keywords Human resource analytics · Strategic human resource management · Capacity building · Ethics. · Ethical leadership

---

R. Afedzie (✉)
Department of Management Studies, Pentecost University, Accra, Ghana
e-mail: rafedzie@pentvars.edu.gh

E. O. Kwakye
Department of Management Studies, Koforidua Technical University, Koforidua, Ghana
e-mail: eugene.okyerekwakye@ktu.edu.gh

M. Dzingirai and S. O. Atiku (eds.), *Strategic Human Resource Analytics and Intelligent Data-Driven Systems*, https://doi.org/10.1007/978-981-92-2623-8_7

1 Introduction

Over the past 20 years, data analytics has become an important part of any business strategy. It has changed how companies hire people, use their resources, and make decisions. Data-driven insights have improved the accuracy of predictions and the efficiency of strategy in many areas, such as marketing, finance, and operations (Davenport & Dyche, 2013; Tambe et al., 2019). Human Resource Analytics (HRA), commonly referred to as people analytics, has become a prominent scientific approach for understanding employee behavior and improving organizational efficiency. HRA helps people make decisions based on facts in areas like hiring, performance reviews, succession planning, and talent management by systematically gathering and analyzing employee data (Boudreau & Cascio, 2017; Huselid, 2018). This field has advanced significantly through the incorporation of Artificial Intelligence (AI), Machine Learning (ML), and predictive algorithms, enabling organizations to pinpoint potential leaders, anticipate training requirements, and project employee turnover with unprecedented accuracy (Dahlbom et al., 2020).

Even though technology has come a long way, the ethical issues surrounding HR analytics have gotten a lot of attention. As companies gather a lot of information about their employees, basic issues arise about privacy, consent, openness, fairness, and the moral limits of using data. Algorithms that seem very fair may still discriminate if they are trained on biased datasets, which would make inequities worse instead of better (Manroop et al., 2024; Tursunbayeva et al., 2021). As a result, the conflict between individual rights and administrative efficiency has become a major topic in modern HR analytics. Data ownership is another issue because employees often do not know how their data is used for management, forecasting, or have much say in how it is used. Our twenty-first-century HR professionals and scholars concur that the incorporation of ethics into HR analytics is not only advantageous but also imperative for ensuring business sustainability, accountability, and trust. In developed countries, there are strict rules about ethical issues related to HR analytics. This is not the case in Ghana, which is an emerging economy.

Most organizations in Ghana still work in unstable digital ecosystems, even though many of them, especially those in banking, telecoms, and manufacturing, are using HR analytics solutions to make their employees more productive. Problems like bad data governance frameworks, not enough technology infrastructure, and not enough experience with analytics make it hard to implement effectively. In addition, the inconsistent enforcement of the Data Protection Act (2012) makes it unclear who is responsible and who is following the rules when it comes to processing employee data. Due to its high cost and the lack of skilled data specialists, only a small number of big companies can afford to use HR analytics software. Similarly, many public sector organizations still use old systems that are prone to inefficiencies and security breaches. Many organizations do not have the procedures in place to make sure that data is handled responsibly. These contextual restrictions make ethical risks higher, especially when it comes to data privacy, algorithmic bias, and fairness (Mensah,

2022). In addition, Ghana's culture and institutions have an impact on how analytics-related ethics are governed. People may not want to question or criticize analytics-driven decisions in companies with hierarchical cultures, even if the results seem unfair or wrong (Ofori, 2022). Because HR professionals do not usually know about ethical data standards, analytics results are often accepted without being checked. HR analytics can be a good way to improve performance, but they can also hurt employee trust and the company's reputation if ethical protections are not put in place (Chinyamurindi & Louw, 2021). As a result, the Ghanaian situation is a unique and little-studied case for looking into how companies might combine technical innovation with moral responsibility when it comes to managing labor data.

2 Problem Statement

HR analytics has become an important part of managing human capital strategically around the world, but we do not know much about how ethical issues are built into its development and use in Ghanaian businesses. The literature currently available predominantly reflects the experiences of developed economies, characterized by advanced institutional ethics, digital infrastructure, and regulatory frameworks. Ghana, on the other hand, has its own problems because it does not have enough skilled workers, does not enforce data protection laws well, has physical limitations, and has cultural norms that put administrative authority above individual privacy. These factors create the potential for unethical conduct, encompassing algorithmic bias, unauthorized data access, and ambiguous decision-making processes that undermine justice and transparency. Even though the Data Protection Act (2012) is in place, employees are still at risk of having their personal information stolen because the law is not enforced strictly enough and there are no clear rules for HR analytics. This empirical knowledge gap raises significant enquiries, such as the significance of ethics in HR analytics for organizations in Ghana. What internal policies or frameworks govern the collection, analysis, and utilization of employee data? In addition, how much privacy and rights of employees are protected during this process? Addressing these concerns is crucial for safeguarding the benefits of analytics while maintaining organizational integrity and human dignity, and for aligning Ghanaian HR practices with global ethical standards.

3 Research Questions

- How do private and public sector organizations in Ghana employ HR analytics in their work and its ethical principles thereof?
- What are some of the key benefits derived from HR analytics by private and public organizations in Ghana?

- What are some of the primary ethical challenges associated with HR analytics implementation in Ghana?

This study theoretically enhances the understanding of ethics in human resource analytics through the application of stakeholder theory, fairness theory, and the Resource-Based View (RBV). The models stress that ethical data governance builds trust, protects workers' rights, and keeps business performance high. By situating these theories within the Ghanaian context, the study illustrates how ethical analytics can function as a strategic asset for the sustainable development of human capital. This bridges the disparity between global ethical frameworks and local realities. It gives a full picture of how technical progress, cultural norms, and institutional capacity affect ethical decision-making in HR analytics.

The chapter provides Ghanaian regulators, legislators, and HR professionals with valuable insights. It highlights institutional shortcomings in the regulation of ethical data and proposes measures to improve compliance with Act 843 and increase accountability in HR systems. The Ministry of Employment and Labor Relations and the Data Protection Commission can also use the study to help shape national policy. It shows how HR analytics can help Ghana reach its broader goals for digital transformation.

4 Literature Review

The literature review will examine significant studies conducted in the last ten years regarding HR analytics. The discussion on the literature will also address some of the gaps in the literature. The objective is to rectify certain inadequacies present in the literature concerning public and private sector organizations in Ghana.

4.1 Concept and Evolution of Human Resource Analytics

HR analytics is the use of data analysis methods to make human resources management better. This process helps improve HR functions like hiring, performance reviews, employee engagement, and keeping staff by collecting, analyzing, and making sense of relevant HR data (Dubey, 2023; Gupta et al., 2023). At first, HR analytics was all about using data on how well employees performed their jobs to make the administration run more smoothly. However, the field has changed a lot in the last decade. As big data, machine learning, and artificial intelligence have grown, HR analytics has gotten more advanced. It now has more advanced predictive and prescriptive tools that help companies make proactive and data-driven HR decisions (Awasthi et al., 2023; Dubey, 2023).

Advancements in new technology have enhanced human resources from a reactive administrative role to a strategic decision-making partner. Data-driven insights now guide hiring, performance reviews, and succession planning. Several businesses are using predictive models to find top employees, predict how many employees will leave the organization, and see how well training programs are working.

Companies like the Social Security and National Insurance Trust (SSNIT), Ghana Commercial Bank (GCB), and MTN Ghana have begun using analytics in HR operations in Ghana, particularly for workforce planning and performance management (Data Protection Commission, 2024; PwC Africa, 2023). These new ideas show that Ghana is moving toward evidence-based HRM, even though there are still ethical and practical problems, such as data protection, a lack of analytical skills, and following the rules.

4.2 Theoretical Foundations of HR Analytics

4.2.1 Stakeholder Theory

Freeman's (1984) Stakeholder Theory posits that corporations possess strategic and ethical obligations to all stakeholders affected by their decisions, rather than solely to shareholders. The theory says that employees, whose personal and behavioral data are studied, are important stakeholders whose rights and welfare must be protected when human resource analytics (HRA) is used. For HR analytics to be ethical, data collection, interpretation, and use must be open, informed, and fair.

Stakeholder Theory is particularly relevant in Ghana due to hierarchical organizational cultures and a profound respect for authority, which often limits employees' ability to contest data-driven decisions (Asiedu-Appiah et al., 2021). ntegrating stakeholder principles can help balance power dynamics by making sure that all staff members are involved in discussions about data privacy and ethics. This idea is especially important in Ghana, where most workplaces are collectivist and relationships and trust are still important parts of organizational life (Ofori, 2022). Using stakeholder thinking, you can make sure that HR analytics is used as a collaborative effort based on fairness and openness, not as a way to keep an eye on things. The Ghana Data Protection Act (2012, Act 843) also reflects the values of stakeholders by requiring responsibility and justice in the handling of personal data.

4.2.2 Ethical Decision-Making Theory

The Ethical Decision-Making Theory (Rest, 1986; Treviño, 1986) is an important way to look at how people and groups make moral choices when they have many data. The theory posits that moral awareness, judgment, intention, and action constitute the components of ethical activity. These steps in HR analytics are like figuring out the moral implications (like bias or discrimination), weighing the ethical risks (like data misuse), deciding on the best course of action, and putting responsible policies into place.

This theoretical framework makes it easier to understand how HR professionals deal with moral dilemmas when they use analytics tools. HR professionals in Ghana often confront the ethical dilemma between the obligation to safeguard employee privacy and the imperative for managerial efficiency. Appiahene et al. (2022) assert that the absence of formal ethics training in numerous Ghanaian HR departments hinders practitioners' ability to accurately evaluate the moral ramifications of analytics-driven decisions. To prevent inadvertent harm, Ethical Decision-Making Theory underscores the importance of institutional support and ethical proficiency. Cultural factors also play a role in moral reasoning in Ghanaian businesses. According to Mangaliso (2020) and Metz (2018), African ethics often prioritize social harmony (Ubuntu) over individual sovereignty. This indicates that ethical awareness can be communicated through group discourse rather than solely through individual conscience. Ethical Decision-Making Theory goes beyond its Western roots by adding an Afrocentric moral point of view. This helps us better understand how ethics work in groups.

4.3 Ethical Considerations in Human Resource Analytics

A set of rules that tells people what is right and wrong is called ethics (Resnik, 2024). Ethics is not something you are born with; you learn it through socialization, education, and experience. It includes both understanding things cognitively and feeling them emotionally. Professional ethics necessitate adherence to moral principles, continuous education, and the pragmatic implementation of ethical knowledge (Taquette et al., 2022). HR analytics, which uses data-driven insights to improve how people are managed at work, has changed how human resources is done today. The greater use of analytics in HR work has brought up four related ethical risks: bias and discrimination, privacy, trust, and openness.

4.3.1 Bias and Discrimination

Algorithmic bias is one of the most important moral problems in HR analytics. Algorithms trained on historical datasets can reproduce systemic disparities rooted in socioeconomic status, gender, or ethnicity (Tambe et al., 2019). This issue is especially important in Africa, where uneven statistics and a lack of data often show bigger socio-economic differences. In Ghana, gendered employment trends and the lack of women in leadership and STEM roles make this problem worse (Ghana Statistical Service, 2024; ILO Africa, 2023). When HRA systems use biased datasets, they could make inequality worse instead of better. To fight this, organizations must use a variety of training datasets, do algorithmic audits often, and make sure that people from different fields supervise analytical procedures.

4.3.2 Transparency

Transparency is how clearly businesses tell their employees how they collect and use their data. If analytics systems are not clear about what they do, employees may not trust them and may even fight against them. For HR analytics to be ethical, it is important to be open about the goal, scope, and possible outcomes of data collection (Jobin et al., 2019). The UNESCO Recommendation on the Ethics of Artificial Intelligence (2021) and the Organization for Economic Co-operation and Development (OECD, 2021) both mention that moral data-driven systems need to be clear and have human oversight. AI governance is still very new in Ghana, and many HR departments do not have official data ethics committees or clear ways to hold people accountable. Creating institutional data ethics boards made up of HR, legal, and IT professionals can help hold people accountable within the organization and make sure that everyone follows national and international ethical standards. The institutionalization of ethical oversight is a crucial initial step in Ghana's implementation of ethical HR analytics.

4.3.3 Trust

Trust is the basis of ethical HR analytics. When employees are sure that their data will be handled fairly, privately, and with integrity, they are more likely to work with analytics projects (Bhandari, 2024). Conversely, unethical data practices that adversely affect organizational culture and morale encompass clandestine surveillance and ambiguous decision-making. The cultural context also affects how people think about ethics. In Ghana, collectivist cultural orientations prioritize interpersonal trust, connections, and respect over adherence to formalities (Ofori & Laryea, 2021). Therefore, cultural sensitivity and a mix of relational engagement and technical compliance are important parts of ethical HRA. To build trust and shared responsibility for moral outcomes, organizations should actively involve their employees in the design of analytics systems.

4.4 Empirical Perspectives

Although empirical data from Africa remains limited, it is increasing and illustrates the continent's gradual adoption of digital HR tools. Studies conducted in South Africa, Nigeria, and Kenya (Chinyamurindi & Louw, 2021; Muriithi, 2022) indicate that organizations recognize the strategic value of analytics; however, ethical and infrastructural challenges persist. There are a lot of problems that are very similar to the ones in Ghana, such as not being able to read data, not having enough IT infrastructure, and not enforcing rules strictly enough.

In Ghana, the use of HR analytics has been mixed. Most of the projects are in big, formal-sector companies. A recent report from the Ghana Data Protection

Commission (2023) expresses that only 36% of registered organizations had data protection officers, and less than 20% checked to see if HR data was being used in an ethical way. This gap shows that institutions do not have enough resources to use ethical analytics methods, especially in small businesses and public organizations.

According to local research by Ohemeng and Adusah-Karikari (2021), public sector organizations in Ghana often do not have clear ethical rules for digital HR management. This leads to messy and random ways of using data. Danso and Boateng (2023) found that Ghanaian HR managers are very worried about privacy violations and not having enough consent processes, even though they know that analytics can help with talent management and performance monitoring.

Cultural issues also have a big effect on the ethical environment of HR analytics in Ghana. Collective ideals and respect for authority often stop employees from questioning data-gathering methods, which leads to the passive acceptance of organizational monitoring. This situation shows how important it is to have ethical frameworks that take into account the situation and combine technical protections with programs for education and empowerment.

Policy-level efforts have been made to close these gaps. The Ghana Data Protection Commission has issued a number of public warnings saying that employers must say why they are collecting data, limit access to only authorized people, and keep employee data for as long as it is needed. The National Digitalization Policy (2022) also encourages the careful use of data analytics in the public sector to keep things open and accountable. However, these rules have not been fully turned into legally binding HR analytics requirements yet.

5 Methodology

5.1 Research Design

The study employed a multi-site qualitative multiple-case research design. This study analyzed multiple organizational contexts to compare and contrast the manifestation of ethical dilemmas in Human Resource Administration across diverse settings, rather than employing a single-case design focused on a singular constrained unit. This approach is in line with Yin's (2018) definition of multiple-case research, which allows for cross-case validation and replication reasoning. The study examined HR analytics practices in twenty (20) organizations within Ghana's public and commercial sectors to analyze variations in institutional frameworks, HR maturity, and analytics adoption. A multi-site strategy is advantageous as it facilitates comparative cross-case research and enhances the transferability of findings by intentionally sampling organizations with diverse sizes, sectoral roles, and technological complexities.

5.2 Sampling Strategy and Sample Size

Participants were selected through purposive (criterion-based) sampling, ensuring direct involvement in HR analytics and ethical oversight. As stated by Saunders et al., (2016, p. 485), this sampling method enables the researcher to utilize particular judgments or criteria in selecting individuals who are most pertinent to the research objectives. Participants were required to (a) hold a senior position in HR, HRIS, or compliance, (b) possess a minimum of three years of professional experience, and (c) demonstrate familiarity with the collection, utilization, or regulation of HR data and (d) willingness to participate and give informed consent.

The study's target group came from both the public and private sectors. The Ministry of Finance, the Ministry of Health, the Ghana Education Service (GES), the Electoral Commission of Ghana, the Ghana Armed Forces (GAF), the Environmental Protection Agency (EPA), the Ghana Revenue Authority (GRA), the Ghana Cocoa Board (COCOBOD), the Bank of Ghana (BoG), and the Ghana Police Service were all part of the public sector. The private sector organizations that made up the population were Unilever Ghana Ltd, MTN Ghana, Melcom Group, Nestle Ltd, ABSA Bank, Multimedia Group Ltd, Jospong Group of Companies, Ghana Oil Company Ltd (GOIL), Kasapreko Company Limited, and Tullow Group.

This chapter sought to highlight instances of ethical concerns in HR analytics. As such, participants who have worked in the selected organizations for five years provided rich data on the subject matter. Three participants for the semi-structured interview process were selected from the HR department of each organization. Overall, forty participants were interviewed from the total of sixty selected. The forty participants were based on the point of saturation, where the forty-first participant provided no new information. The point of saturation in qualitative studies aligns with a sample size between 30 and 50 interviews, based on the research objectives and diversity of participants.

5.3 Determining and Reaching Thematic Saturation

Thematic saturation was identified through iterative monitoring. The research team looked at the codes that were changing and checked to find out if new information was still coming in after every five interviews. Saturation was considered attained when three consecutive interviews produced no novel primary or secondary codes. This methodical process confirmed the adequacy of depth and breadth by ensuring that data collection ceased only when the analytical categories were fully developed and no new themes were arising (Saunders et al., 2018). By the 38th interview, codes had become repetitive, reinforcing established patterns instead of offering fresh concepts. Two final confirmatory interviews were conducted to ensure completeness. This strict method gave us confidence that we had really reached thematic saturation.

5.4 Data Collection

Data were collected via semi-structured, face-to-face interviews conducted at participants' workplaces between October 2024 and March 2025. An interview outline was utilized at all locations to ensure uniformity while enabling participants to expand on emerging subjects. The main areas that were looked at were: (i) how and how much HR analytics is used in the company; (ii) data sources and governance methods; (iii) the pros and cons of analytics; (iv) ethical rules, permission procedures, and oversight mechanisms; and (v) suggested protections and capacity gaps. Some example questions were: "Tell us how your company collects and uses employee data to make HR decisions." "How do employees find out how data is used?" In addition, "What internal checks are in place to make sure that analytics don't lead to biased or unfair results?" The appendix has the whole interview protocol.

After each interview, a structured observation template was used to carefully write down field notes and non-verbal cues like body language, mood, and hesitation. These observations were not coded as independent "body language" codes unless they directly supported an interpretive assertion; instead, they were treated as supplementary data and integrated into the analysis through memo and triangulation with transcript codes (for instance, a pronounced emotional response to a policy that elucidated a theme of mistrust). All interviews were recorded on audio and transcribed word-for-word with the participants' permission. The research team was the only one that could access the encrypted drives where the anonymized transcripts were safely stored.

5.5 Data Analysis: Theme Analysis with Transparency and Rigor

The data were examined through thematic analysis following Braun and Clarke's (2006) six-step methodology. We used NVivo (version 12) to organize transcripts, codes, and memos to make them more open and reproducible.

Generating initial codes: NVivo employed an open coding methodology; two researchers established a preliminary codebook and independently coded the initial ten transcripts. Codes encoded semantic content ("consent procedures," "algorithmic bias") and interpretive assessments ("managerial defensiveness").

Looking for themes: We used iterative team workshops to sort the first codes into possible themes, look at trends, and make tentative thematic maps.

Checking themes: Candidate themes were compared to the dataset in NVivo to make sure they were internally consistent and externally diverse. Codes were moved around or combined as needed.

Defining and naming themes: The final themes were articulated through precise definitions and illustrative quotes; sub-themes were created to encapsulate nuance. To make the report, the themes were connected to the study topics, ethical theory, and Ghanaian policy environment to give analytical interpretations.

Inter-coder reliability and reflexivity: To enhance confidence, two researchers independently coded the initial subset of transcripts and subsequently compared their coding applications. We looked at coding problems in weekly meetings and came to an agreement on how to fix them. If there was still disagreement, a senior third researcher was asked for advice. A living codebook had definitions for codes and rules for making decisions. New codes could only be added with the team's permission. Qualitative analysis values interpretive depth over numerical reliability, yet the inter-coder method enhanced consistency and diminished idiosyncratic coding. Cross-checking 25% of coded transcripts between two researchers enhanced inter-coder reliability, with discrepancies addressed through discourse rather than statistical metrics, aligning with interpretive qualitative methodologies (O'Connor & Joffe, 2020). The repeated comparisons fostered interpretive depth and analytical coherence rather than mere numerical concordance.

The study employed the four established criteria of trustworthiness, credibility, dependability, confirmability, and transferability, as proposed by Lincoln and Guba (1985), to uphold methodological rigor. Long-term fieldwork, member validation, and the triangulation of data sources (interviews, field notes, and policy texts) all enhanced credibility. To guarantee accuracy, selected participants were requested to review summarized interpretations of their statements (Nowell et al., 2017).

An audit trail that kept track of research choices, codebook creation, and changes to analytical frameworks helped to show that the results were reliable. All changes to coding or theme names were written down in NVivo notes. Researchers kept reflexive journals to write down their assumptions, biases, and positions during the analysis. This helped to support confirmability. The detailed description of the research environment facilitated transferability and enabled readers to evaluate the applicability of the findings to various African contexts.

The research adhered to rigorous ethical standards. The people who took part in the study were properly told what it was about, and their privacy, anonymity, and choice to take part were all protected. Using pseudonyms and organizational codes helped protect people's identities. The rules for keeping data safe were in line with Ghana's Data Protection Act, 2012 (Act 843). The encrypted digital recordings and transcripts were kept on devices that only the researcher could access. The notes and consent papers were kept in secure cabinets. All computer files were encrypted and kept safe, and the field data will be destroyed when the research is done.

6 Findings and Discussions

This section summarized the main points made by the study participants about how to use Human Resource Analytics (HRA) ethically in some public and private organizations in Ghana. Based on the views of 40 HR professionals with different ages, genders, levels of education, and years of experience (see Table 1), the results show real-world ethical issues, practical problems, and perceived benefits of using HRA.

Ethical violations in analytics are one of the foremost concerns confronting organizations today.

Below are key categories and instances of such breaches:

6.1 Theme 1: Algorithmic Bias and Fairness

One of the biggest moral problems with using Human Resource Analytics (HRA) in Ghanaian businesses has been the problem of fairness and bias in algorithms. Participants reported that algorithmic tools often reproduced existing disparities, particularly in hiring, promoting, and evaluating employees, despite analytics systems being marketed as objective and data-driven. This study underscores the growing concern that technology, despite its apparent impartiality, may perpetuate structural and historical biases inherent in the analyzed data (Dastin, 2018; Raghavan et al., 2020).

In Ghana, algorithmic bias in HR decision-making shows up as small but important changes. Participant 5, an HR Manager in the Private Sector, said, "When we started using analytics to narrow down applicants, we noticed that fewer women and older candidates were being chosen." One of the main things the algorithm looked for was a work history that did not have any gaps, which hurt women who took parental leave. This experience aligns with global findings indicating that biased data inputs often result in discriminatory outputs, particularly concerning under-represented or marginalized populations (Newman et al., 2020). However, in Ghana, these biases are made worse by bad data quality, incomplete personnel records, and a lack of digital supervision.

Participant 8 (Public Sector HR Officer) also said the following: "The system likes employees who keep their digital records up to date." Since the analytics platform cannot see all of our data, those of us who still keep manual files in our departments often get lower ratings. This comment highlights the disparities in infrastructure that hinder fair algorithmic evaluation in Ghana's public institutions, where both digital and paper data systems are used. Even when algorithms are not inherently biased, deficiencies in digitization and incomplete data collection can lead to unintentional prejudice, as demonstrated by Boateng et al. (2020), Ohemeng et al. (2021).

In theory, the fairness challenge in HR analytics aligns with Stakeholder Theory, which posits that businesses possess a moral obligation to consider the rights and welfare of all affected parties (Freeman, 1984). When algorithmic systems favor some

Table 1 Demographic profile of respondents

Respondent	Gender	Age	Education	Job position	Years of experience	Public institution
P1	Male	Over 46	MSc Human Resource Management	Director of HR Policy	25	Ministry of Finance
P2	Female	36–45	MSc Public Health Administration	HR & Data Compliance Officer	18	Ministry of Health
P3	Male	36–45	MA Education Planning	HR Analytics Lead	20	Ghana Education Service (GES)
P4	Female	Over 46	MSc Human Resource Law	Ethical Standards Officer	22	Electoral Commission of Ghana
P5	Male	Over 46	MSc Organizational Security	HRIS Specialist	23	Ghana Armed Forces (GAF)
P6	Female	36–45	MSc Environmental Policy & HRM	HR Data Coordinator	16	Environmental Protection Agency (EPA)
P7	Male	26–35	MBA Human Resources	HR Business Analyst	10	Ghana Revenue Authority (GRA)
P8	Male	Over 46	MPhil Public Administration	HR Technology Integration Officer	19	Ghana Cocoa Board (COCOBOD)
P9	Female	26–35	MSc Accounting & HR Analytics	HR Data Governance Officer	11	Bank of Ghana (BoG)
P10	Male	26–35	MSc Industrial Psychology	HR Metrics Analyst	8	Ghana Police Service
P11	Male	Over 46	MBA Human Resource Strategy	Group Head of Human Capital	21	Unilever Ghana Ltd
P12	Female	36–45	MSc Data Science for HRM	HR Analytics Manager	15	MTN Ghana
P13	Male	36–45	MSc Workforce Analytics	Talent Analytics Lead	14	Melcom Group
P14	Female	26–35	MSc Organizational Development	People and Culture Data Officer	10	Nestle Ltd

(continued)

Table 1 (continued)

Respondent	Gender	Age	Education	Job position	Years of experience	Public institution
P15	Male	36–45	MSc Strategic HR Management	Compensation & Analytics Specialist	13	Absa Bank
P16	Male	Over 46	MSc Communication & HR Ethics	Head of People Analytics	19	Multimedia Group Ltd
P17	Female	26–35	MSc HR Information Systems	HR Tech & Ethics Officer	9	Jospong Group of Companies
P18	Male	26–35	MSc HR Analytics & Policy	Workforce Planning Analyst	7	Ghana Oil Company Ltd. (GOIL)
P19	Female	36–45	MSc International HRM	HR Ethics and Compliance Manager	12	Kasapreko Company Limited
P20	Male	26–35	MSc Applied HR Analytics	Senior HR Data Analyst	10	Tullow Group

types of staff over others, they go against the stakeholder idea of treating everyone the same. The Ethical Decision-Making Theory (Rest, 1986) further substantiates this notion, suggesting that managers may recognize prejudice yet refrain from intervention due to insufficient technical proficiency, institutional support, or ethical consciousness. In Ghana, decision-makers might neglect their moral duties in favor of technological experts, as HR departments often outsource analytics to external companies.

Socio-Technical Systems Theory explains how the differences between the culture of an organization and its technical systems cause these ethical problems. Companies in Ghana often use analytics tools without making sure they work well in their area or that people are watching the machines to make sure they get better results. Participant 14 (IT and Data Systems Officer, Private Telecommunications Firm) said that most of the algorithms we use are pre-set from outside sources. We do not know enough to say if they fit our data or our workforce. This dependence reveals a socio-technical disparity, as imported methodologies are employed without rigorous evaluation for equity within Ghana's distinct data context.

Despite these concerns, there are more and more efforts in Ghana to reduce prejudice and make algorithms more accountable. Some big companies, especially those in the banking and telecom industries, have started regularly checking their HR analytics outputs to look for patterns of unfairness. Others have started teaching HR departments how to properly look at analytical results and question automated suggestions. These changes indicate a gradual transition toward "ethical analytics maturity," as delineated by Sharma and Kaur (2023, p. 23), wherein fairness is deliberately integrated into system evaluation rather than assumed.

6.2 Theme 2: Data Privacy and Surveillance

Data privacy and surveillance have been one of the most sensitive ethical issues in the use of Human Resource Analytics (HRA) in Ghanaian businesses. Participants expressed apprehensions regarding the magnitude of behavioral and personal data collection, retention, and analysis, frequently conducted without transparent objectives or explicit consent, despite analytics yielding valuable insights for improving employee performance, recruitment, and retention. These findings underscore an ethical dilemma between employee privacy and organizational efficiency, increasingly observed in both African and global contexts (Manroop et al., 2024; Minbaeva, 2021). Participant 12, an HR officer at a public health institution, said, "We collect biometric data and keep track of staff attendance, but many employees don't know how this data is used or who has access to it." This lack of openness shows how privacy breaches are more likely to happen because of poor data governance and communication than on purpose. Participant 21 (Private Telecommunications Firm) said something similar: "The system keeps track of logins, emails, and even time spent on the screen." Some workers see it as spying on them instead of a way to judge their work. This feeling of being watched all the time has made employees more distrustful and anxious. Fuchs (2017, p. 81) calls this "digital surveillance capitalism," which is a work environment where monitoring is used instead of communication as a management strategy.

From a theoretical standpoint, Stakeholder Theory (Freeman, 1984) underscores that workers are legitimate stakeholders entitled to privacy, transparency, and informed consent, rather than merely data subjects. According to the Ethical Decision-Making Theory (Rest, 1986), managers who justify intrusive surveillance as "necessary efficiency" without thinking about its moral implications are also doing something wrong. In Ghanaian companies, where cultural norms value respect for authority and job security, workers hardly ever question these kinds of things. This keeps the power imbalance in the use of data going. In Ghana, a lack of institutional capacity and enforcement of rules makes it more likely that employees' data will be used in unethical ways. The Data Protection Act (2012, Act 843) gives people a legal way to protect their personal data, but many businesses do not have data protection officers or procedures for following the law. A report from the Ghana Data Protection Commission (2023) says that less than half of registered businesses fully follow the Act's rules, especially those about limiting data, getting consent, and limiting purpose. HR analytics procedures often operate in ambiguous domains where employee data is collected in excess of intended parameters, complicating the differentiation between analytics and surveillance.

6.3 Theme 3: Employee Autonomy, Consent, and Transparency

Three interconnected ethical concerns, employee autonomy, consent, and transparency, have surfaced as determinants of the perception and implementation of HRA in Ghanaian enterprises. Participants frequently noted that employees are seldom consulted or informed regarding the collection, analysis, or utilization of their data in decision-making processes, despite HRA's assertions of delivering evidence-based insights for talent management. Echoing international findings that ethical analytics must be based on informed consent and participatory governance, this limited participation jeopardizes autonomy and erodes organizational trust (Bodie et al., 2016; Manroop et al., 2024). Participant 17 (Public Sector HR Specialist) said, "We collect employee data for analytics, but employees are never told how it affects promotions or evaluations." This lack of communication hurts employee agency and turns analytics into a tool for control instead of empowerment. Management says it is system-based when workers ask why certain choices are made. Participant 26, an HR Manager at a Private Bank, said, "That makes people less likely to ask questions." This lack of clarity shows how much managers are relying on technology to explain their decisions, often at the cost of communication and accountability between people. Stakeholder Theory (Freeman, 1984) offers a valuable ethical perspective, asserting that employees ought to be regarded as moral agents entitled to information and participation, rather than mere data inputs. The Ethical Decision-Making Theory (Rest, 1986) backs this up by saying that moral courage, awareness, and judgment are all important for ethical behavior and that HR departments need to have these traits ingrained in them. Organizations that do not value openness risk breaking psychological contracts and losing employee loyalty because decisions seem to come from outside the organization and cannot be questioned.

6.4 Theme 4: Profiling, Behavioral Modeling, and Trust

Another important ethical part of using Human Resource Analytics (HRA) in Ghanaian businesses was figuring out how to build trust, model behavior, and profile people. Participants expressed heightened apprehensions regarding the potential for misinterpretation, stereotyping, and the erosion of workplace trust as HR systems increasingly employ predictive algorithms to assess employee performance, retention, and behavioral tendencies. When organizational cultures prioritize managerial control over participative ethics, these moral dilemmas are intensified (Ajunwa, 2020; Manroop et al., 2024).

Participant 10, the Public Service HR Manager, said, "The analytics system classifies staff as 'high risk' or 'low performers' based on attendance or past disciplinary actions." Once the label has been put on, it is hard to take it off. This shows how static profiling can be bad because predictive data can make workers look bad and change managers' minds even when behavior gets better. Recent studies have shown similar outcomes, indicating that algorithmic profiling may exacerbate bias and diminish fairness in hiring practices (Newman et al., 2020; Raghavan et al., 2020; Tambe et al., 2019). In Ghana, inconsistent data entry and insufficient contextual calibration heighten the risk of behavioral misclassification. The datasets used to calibrate many analytics systems in the public sector are based on Western corporate environments and do not reflect Ghana's sociocultural work norms (Appiahene et al., 2022; Boateng et al., 2020). Participant 28, a private telecommunications analyst, said, "We found that our algorithm kept flagging employees who questioned supervisors as difficult." The system sees speaking up as a bad thing, though, because in our culture, it is often seen as rude. This example shows how algorithms that come from other cultures might misread culturally subtle behaviors, making moral dilemmas out of differences in context. Socio-Technical Systems Theory (Trist & Emery, 1973) provides a theoretical framework for understanding these ethical conflicts. The idea is that ethical and effective technical systems must be compatible with the institutional, social, and cultural contexts in which they operate. In Ghana, workers see analytics as a way to keep an eye on things rather than a way to help, which causes distrust because imported technology does not always follow local ethical standards. Fairness Theory (Adams, 1965) posits that perceived unfairness in the application of profile results may lead to psychological withdrawal, diminished morale, and disengagement. The research conducted by Minbaeva (2021) and Sharma and Kaur (2023) corroborates these findings.

6.5 Theme 5: Organizational Culture, Trust, and Ethical Governance

In Ghanaian organizations, the main things that affect how people see, use, and keep Human Resource Analytics (HRA) are the culture of the organization, trust, and ethical governance. The research revealed that governance frameworks, leadership integrity, and the values that dictate the use of ethical analytics are equally critical to its success or failure as technology. Participants repeatedly stressed that even advanced analytics tools can be misused, ignored, or not trusted by staff members if there is no culture of ethics.

Participant 9 (HR Executive, Manufacturing Sector) said, "Our analytics software is available, but management only uses it when it's convenient for them." Sometimes, people ignore findings that do not match what they want. This phrase describes a common problem in Ghanaian organizations: decisions are often made based on hierarchy and individual influence rather than evidence-based recommendations, which leads to inconsistent decision-making. According to Ohemeng and Adusah-Karikari (2021), Ghana's institutional culture is still very hierarchical, which makes it hard to make sure that everyone is responsible and open at every level of government. Many organizations have trouble with ethical governance because there are no official rules or groups to make sure that data is used ethically. Participant 13, the HR Director for the Public Sector, said, "We have HR data systems, but no one is in charge of making sure the analytics are ethical." People think that IT will handle it. This lack of clear ownership over decisions involving data is what Floridi (2018, p. 79) calls "the ethics gap." It shows how accountability is spread out. Because the HR and IT departments do not work together, ethical governance breaks down. Stakeholder Theory (Freeman, 1984) gives a good reason for this problem. The theory says that companies have a moral duty to consider the needs of everyone involved, not just management. However, in many Ghanaian companies, employees are not part of ethical discussions because decisions about how to use data and analytics are still made at the highest level. This makes people less responsible and less sure of themselves. The Ethical Decision-Making Theory (Rest, 1986) also helps to explain these results. This theory says that moral behavior in businesses depends on knowledge, discernment, and motivation, all of which need to be developed through leadership development and institutional culture. According to Participant 35 (HR Manager, Telecommunications Sector), employees are more open to analytics when management explains the process and admits mistakes. When leaders set a good example, people trust them more. This reaction shows how important it is to have moral communication and consistent leadership in order to build trust in analytics. Sharma and Kaur (2023) and Minbaeva (2021) have done real-world studies that back this up. They show that when leaders act ethically and talk to their staff openly, trust in data systems goes up.

To fix these problems with governance, some Ghanaian organizations are starting to make data protection committees, staff training programs, and ethical review boards permanent. These actions show that people are starting to understand how important accountability is in HR systems. They also follow international best practices for ethical AI governance (Ghana Data Protection Commission, 2023; Manroop et al., 2024). However, because these efforts are spread out and mostly focused on global companies, many local businesses do not have official ways to keep an eye on things.

7 Recommendation

- **Phase the Implementation of Ethical Audits**

Instead of putting in full algorithmic audit trails all at once, organizations should do it in stages. First, basic bias detection and regular data-quality checks should be done. Then, more in-depth algorithmic audits should be undertaken. This systematic plan helps build a culture of accountability over time while making sure that it works within Ghana's limited resources and infrastructure.

- **Strengthen Institutional and Policy Frameworks**

The Ghana Data Protection Commission (DPC) and the Ministry of Employment and Labor Relations should set clear rules for how to use HR analytics in an ethical way. These criteria should include fairness checks that are required, procedures for getting employee consent, and data protection audits that follow both national laws and best practices around the world.

- **Build Local Capacity for Ethical Data Management**

Academic institutions and professional HR associations, such as the Institute of Human Resource Management Practitioners (IHRMP-Ghana), should develop certification programs in data governance, algorithmic transparency, and HR analytics ethics. These kinds of programs will help close the current skills gap and give professionals the tools they need to make ethical choices without relying too much on outside suppliers.

- **Promote Organizational Transparency and Employee Inclusion**

HR departments should make internal ethics committees that include employees. These committees can look at the rules for analytics, deal with complaints, and hold training sessions on how to protect data and use it in an ethical way. Getting employees involved builds trust, lowers suspicion, and encourages a sense of shared moral duty.

- **Embed Ethical Culture into Leadership Practice**

Ethical leadership training should be a part of executive development programs in both the public and private sectors. Leaders who are open and honest can reduce resistance that comes from differences in information or culture and build trust in the use of analytics.

8 Avenue for Further Research

This study's exploratory, qualitative design provides significant contextual insights; however, it also presents limitations that suggest new avenues for research. Generalizability is constrained by the small, deliberately selected sample of forty participants from twenty organizations, as experiences may differ by industry or geographic location. Consequently, to elucidate the evolving trends in HR analytics ethics across Africa, subsequent research must expand upon these findings through comparative, quantitative, and longitudinal approaches. Future research may examine the impact of cultural, infrastructural, and legal disparities on the ethical management of HR analytics by comparing Ghana's experience with that of other African countries such as Kenya, Nigeria, or South Africa. Cross-national studies would reveal patterns of regional convergence or divergence in digital ethics.

A second area that needs more research is using quantitative methods to find out how workers feel about consent, privacy, and fairness in workplaces that use analytics. Surveys and statistical modeling could be used to find out how often certain ethical issues come up. These tools could also look at how employee trust in HR analytics systems, leadership ethics, and company culture are related. This would improve the qualitative insights in this study by giving policymakers and HR professionals' useful measurements and evidence that can be used in other situations.

Longitudinal studies ought to be conducted to assess the impact of regulatory reforms or the establishment of ethical standards, such as the evolving guidelines of the DPC, on employee trust and organizational behavior over time. This would provide empirical evidence of the enduring effectiveness of ethical governance instruments.

To offer alternative paradigms grounded in solidarity, human dignity, and collective accountability, further research should examine the impact of African moral frameworks, such as Ubuntu and the Akan ethics of communal responsibility, on global discourse regarding AI and HR analytics ethics. This section shows that the current study is not the end of the road, but rather the beginning of a new path by linking these future research directions to the limitations of the current study's methods. It provides a platform for diverse, comparative, and longitudinal research that can enhance the theoretical understanding and practical implementation of ethical HR analytics across Africa.

9 Conclusion

This study provides a contextualized perspective that distinguishes it from international literature and represents one of the initial comprehensive examinations of the ethical dimensions of HRA in Ghanaian public and commercial entities. The Ghanaian experience illustrates a more complex interplay among institutional capacity, cultural norms, and technological limitations, in contrast to prior research on HR analytics ethics, which predominantly originates from technologically advanced economies with robust regulatory frameworks. The study is unique as it illustrates how ethical deficiencies in HR analytics are intensified by disjointed data systems, inadequate digital infrastructure, and insufficient regulatory oversight. In Ghana, insufficient datasets, limited analytical proficiency, and an overdependence on external vendors present the most significant risks. This contrasts with industrialized economies, where algorithmic bias and data protection are the primary ethical concerns. These situations make it more likely that people will act unethically because of structural problems that make it hard to hold people accountable and be open about what they do, not because of advanced automation. The impact of leadership and culture on moral conduct is unique. In Ghanaian businesses, formal governance mechanisms are often supplanted by interpersonal trust within hierarchical, relationship-centric cultures. This trust-based strategy fosters unity and commitment; however, it inhibits open discourse regarding bias or data exploitation. Consequently, ethics is embedded in societal norms and leadership behavior rather than codified procedures, underscoring the imperative for culturally pertinent ethical governance frameworks. This study also found that indigenous ethical philosophies, such as Ubuntu and Akan ideas of collective accountability, are useful ways to think about digital ethics in African workplaces. These frameworks prioritize human dignity, mutual respect, and collective accountability. African perspectives could enhance international dialogues on responsible analytics by shifting the focus from compliance to compassion. The findings indicate that Ghana's ethical issues with HR analytics are not isolated incidents; instead, they stem from a systemic lack of development in institutional ethics and data governance. A gradual rise in ethical awareness, as seen in the formation of internal data governance teams, compliance units, and proactive steps taken by the Ghana Data Protection Commission, shows that people are becoming more ethically mature.

References

Adams, J. S. (1965). Inequity in social exchange. In L. Berkowitz (Ed.), *Advances in experimental social psychology* (Vol. 2, pp. 267–299). Academic Press. https://doi.org/10.1016/S0065-2601(08)60108-2

Ajunwa, I. (2020). *The quantified worker: Law and technology in the modern workplace*. Cambridge University Press.

Appiahene, P., Amankwah-Amoah, J., & Addai, M. (2022). Ethical challenges and data governance in human resource analytics: Insights from developing economies. *African Journal of Business Ethics, 16*(2), 45–60.

Asiedu-Appiah, F., Adu-Darko, M., & Danquah, J. (2021). Stakeholder participation and ethical leadership in Ghanaian organizations. *Journal of African Business, 22*(3), 311–330.

Awasthi, A., Sharma, D., & Gupta, N. (2023). Advancing human resource analytics through artificial intelligence and machine learning: Opportunities and ethical challenges. *International Journal of Human Resource Studies, 13*(1), 25–44.

Bhandari, R. (2024). Trust and transparency in HR analytics: An ethical imperative for digital workplaces. *Human Resource Development International, 27*(1), 87–103.

Boateng, K., Mensah, P., & Tetteh, G. (2020). Digitization and data ethics in Ghana's public institutions: Challenges and opportunities. *Ghana Journal of Public Administration, 18*(2), 112–134.

Bodie, G. D., Cherry, J. L., McClure, C. M., & Thorson, A. R. (2016). The role of ethical awareness and participation in organizational decision-making: A stakeholder perspective. *Journal of Business Ethics, 137*(1), 135–148.

Boudreau, J. W., & Cascio, W. F. (2017). Human capital analytics: Why we need to move beyond metrics. *People & Strategy, 39*(2), 16–23.

Braun, V., & Clarke, V. (2006). Using thematic analysis in psychology. *Qualitative Research in Psychology, 3*(2), 77–101.

Chinyamurindi, W., & Louw, G. (2021). Human resource analytics and ethics in Africa: A critical perspective. *South African Journal of Human Resource Management, 19*, 1–10.

Danso, E., & Boateng, K. (2023). Ethical governance and digital transformation in Ghanaian HR practices. *African Journal of Management Research, 30*(1), 44–62.
Dastin, J. (2018, October 10). Amazon scraps secret AI recruiting tool that showed bias against women. *Reuters*.
Davenport, T. H., & Dyche, J. (2013). Big data in big companies. *International Institute for Analytics Report*, 1–32.
Data Protection Commission. (2023). *Annual report on data protection and compliance in Ghana (2022/2023)*. Government of Ghana.
Dahlbom, P., Siikanen, N., Saljasalo, P., & Jarvenpaa, M. (2020). Big data and HR analytics in the digital era. *Baltic Journal Management, 15*(1), 120–138.
Data Protection Commission. (2024). *Annual report on data protection and compliance in Ghana (2023/2024)*. Government of Ghana.
Dubey, A. (2023). The evolution of HR analytics: Data-driven decision-making in human resources. *International Journal of Human Capital Studies, 8*(3), 77–95.
Floridi, L. (2018). Soft ethics and the governance of the digital. *Philosophy & Technology, 31*(1), 1–8.
Freeman, R. E. (1984). *Strategic management: A stakeholder approach*. Pitman.
Fuchs, C. (2017). *Social media: A critical introduction* (2nd ed.). Sage.
Ghana Data Protection Act, 2012 (Act 843) (Ghana).
Ghana Statistical Service. (2024). *Labor force survey: Gender and employment statistics 2023*. Author.
Guest, G., Namey, E., & Chen, M. (2020). A simple method to assess and report thematic saturation in qualitative research. *PLoS ONE, 15*(7), Article e0232076.
Gupta, R., Bhatnagar, J., & Awasthi, A. (2023). Data-driven HR: The next frontier of strategic workforce management. *Asia-Pacific Journal of Business Administration, 15*(2), 156–175.
Huselid, M. A. (2018). The science and practice of workforce analytics: Lessons from the field. *Human Resource Management, 57*(3), 679–693.
ILO Africa. (2023). *Women in leadership and employment trends in Sub-Saharan Africa*. International Labor Organization.
Jobin, A., Ienca, M., & Vayena, E. (2019). The global landscape of AI ethics guidelines. *Nature Machine Intelligence, 1*(9), 389–399.
Lincoln, Y. S., & Guba, E. G. (1985). *Naturalistic inquiry*. Sage.
Mangaliso, M. P. (2020). Ubuntu and business ethics in Africa: Integrating moral philosophy and management practice. *Journal of Business Ethics, 167*(4), 775–789.
Manroop, L., Singh, P., & Ezzedeen, S. (2024). Algorithmic management and HR analytics ethics: An integrative review. *Human Resource Management Review, 34*(1), Article 100889.
Mensah, J. K. (2022). Data protection and ethical challenges in Ghanaian HR systems. *African Journal of Management Research, 29*(1), 45–63.
Metz, T. (2018). An African theory of moral status: Ubuntu as a relational ethic. *Philosophical Papers, 47*(1), 23–44.
Minbaeva, D. (2021). Disrupted HR? Human resource management in the digital age. *Human Resource Management Review, 31*(2), Article 100820.
Muriithi, P. (2022). Digital HR analytics adoption and ethical challenges in sub-Saharan Africa: Evidence from South Africa, Kenya and Nigeria. *African Journal of Information Systems, 14*(4), 233–252.
National Digitalization Policy. (2022). *National policy for digital transformation and data governance in Ghana*. Ministry of Communications and Digitalization.
Newman, D. T., Fast, N. J., & Harmon, D. J. (2020). When eliminating bias isn't fair: Algorithmic reductionism and procedural justice in human resource decisions. *Organizational Behavior and Human Decision Processes, 160*, 149–167.
Nowell, L. S., Norris, J. M., White, D. E., & Moules, N. J. (2017). Thematic analysis: Striving to meet the trustworthiness criteria. *International Journal of Qualitative Methods, 16*(1), 1–13.

Nowell, L. S., Norris, J. M., White, D. E., & Moules, N. J. (2019). Achieving saturation in qualitative research: Reflections and recommendations. *International Journal of Qualitative Methods, 18*, 1–9.

O'Connor, C., & Joffe, H. (2020). Intercoder reliability in qualitative research: Debates and practical guidelines. *International Journal of Qualitative Methods, 19*, 1–13.

Ofori, D. (2022). Organizational culture and ethical governance in Ghanaian enterprises. *Ghana Journal of Management Studies, 14*(2), 101–120.

Ofori, D., & Laryea, J. E. (2021). Trust, culture, and organizational ethics in Ghana: A relational approach. *African Journal of Business and Economic Research, 16*(3), 89–108.

Ohemeng, F. L. K., & Adusah-Karikari, A. (2021). Public service reforms, digitization, and ethical governance in Ghana. *International Journal of Public Administration, 44*(11), 932–944.

Organization for Economic Co-operation and Development. (2021). *OECD principles on artificial intelligence*. OECD Publishing.

PwC Africa. (2023). *Data analytics in human resource transformation: Ghana insights report*. PricewaterhouseCoopers Africa.

Raghavan, M., Barocas, S., Kleinberg, J., & Levy, K. (2020). Mitigating bias in algorithmic hiring: Evaluating fairness interventions in practice. *Proceedings of the ACM Conference on Fairness, Accountability, and Transparency*, 469–481.

Resnik, D. B. (2024). *Ethics of science: An introduction* (2nd ed.). Routledge.

Rest, J. R. (1986). *Moral development: Advances in research and theory*. Praeger.

Saunders, B., Sim, J., Kingstone, T., Baker, S., Waterfield, J., Bartlam, B., Burroughs, H., & Jinks, C. (2018). Saturation in qualitative research: Exploring its conceptualization and operationalization. *Quality & Quantity, 52*(4), 1893–1907.

Saunders, M., Lewis, P., & Thornhill, A. (2016). *Research methods for business students* (7th ed.). Pearson Education.

Sharma, S., & Kaur, M. (2023). Ethical analytics maturity and responsible AI in HRM. *Journal of Business Ethics, 182*(1), 55–72.

Taquette, S. R., Oliveira, G. R., & Monteiro, D. L. M. (2022). Professional ethics and moral education in research practice: A contemporary review. *Journal of Bioethical Inquiry, 19*(3), 321–333.

Tambe, P., Cappelli, P., & Yakubovich, V. (2019). Artificial intelligence in human resources management: Challenges and a path forward. *California Management Review, 61*(4), 15–42.

Treviño, L. K. (1986). Ethical decision making in organizations: A person-situation interactionist model. *Academy of Management Review, 11*(3), 601–617.

Trist, E. L., & Emery, F. E. (1973). *Toward a social ecology: Contextual appreciation of the human condition*. Plenum Press.

Tursunbayeva, A., Di Lauro, S., & Pagliari, C. (2021). Ethical challenges of using people analytics in HRM: A systematic review. *Human Resource Management Review, 31*(1), Article 100767.

UNESCO. (2021). *Recommendation on the ethics of artificial intelligence*. UNESCO.

Yin, R. K. (2018). *Case study research and applications: Design and methods* (6th ed.). Sage.

Richard Afedzie is an Associate Professor of Business Management in the Faculty of Business Administration and Dean of the Evening & Weekend School at Pentecost University, Accra, Ghana. He had his PhD education at the University of Texas at Tyler, Texas, United States of America. His research and consulting interests span Human Resource Development, Innovation Management, Strategic Management, Business Sustainability, Organizational Change and Development, Entrepreneurship, Public Sector Management, Disaster Preparedness and Recovery, and Development Planning. He has consulted for numerous professionals and organizations in both the public and private sectors. His teaching portfolio includes courses in Public Sector Management, Entrepreneurship and Strategic Growth, Business Ethics, Public Policy, Emergency and Disaster Management, and Organizational Change and Management.

Eugene Okyere Kwakye is an Associate Professor at the Koforidua Technical University, Ghana. He is currently the Dean of the Faculty of Management and Business Studies. He completed his PhD and master's degree from the Universiti Teknologi Malaysia. He has published articles in reputable journals such as Journal of E-learning and Knowledge Management, Journal of Knowledge Economy, Journal of Knowledge and Innovation Science, Journal of Sustainability and Agribusiness, and attended several international academic conferences. His research interests are in the area of Knowledge Management, Technology Acceptance, Green Management, Cross-Cultural or Diversity Management and Organizational Development. He has been consulted for several international organizations.

Adaptive Talent Acquisition Models for Navigating Change in a Dynamic Environment

K. A. Dhanya, Shikha Malhotra, and H. Sandhya

Abstract In this era of business boom and competition, organizations are finding various ways to entice, identify, and hire world-class talent to meet organizational goals. Traditional talent acquisition models were mostly human-driven, and later companies started using technology as a support for human-driven procedures. Traditional models always suffer from drawbacks like personal preferences, huge expenses, and outdated hiring processes. The advent of Artificial Intelligence (AI), its rapid development, change in expectations about the workforce, and change in skill set requirements, all these factors have contributed to the undergoing expeditious transformation in talent acquisition models. This chapter examines the transformation of talent acquisition models for contributing towards organizational success in a dynamic environment. In a country like India, labour markets are dynamic, and skill set requirements are also ever-changing. Hence, the initial part of the study examines the features of various types of talent acquisition models followed in India and also at the global level. As a progress study exhibits the talent acquisition process with special emphasis on employer branding and Employee Value Proposition (EVP). The study also delves into the impact of technology and automation on talent acquisition practices. The last part of the study examines the challenges of AI in HR practices, especially in the context of the Volatile, Uncertain, Complex, and Ambiguous (VUCA) world and concludes by sketching the road ahead for talent acquisition models, which leverage data-driven insights, technology, and agile recruitment strategies to build resilient talent pipelines that align with evolving business needs.

Keywords Talent acquisition · Employee Value Proposition · Artificial intelligence · Applicant tracking system · Gig economy · Virtual Reality

K. A. Dhanya · S. Malhotra · H. Sandhya (✉)
School of Business Management, Christ University, Bangalore, India
e-mail: sandhya.h@christuniversity.in

K. A. Dhanya
e-mail: dhanya.ka@christuniversity.in

S. Malhotra
e-mail: shikha.malhotra@christuniversity.in

M. Dzingirai and S. O. Atiku (eds.), *Strategic Human Resource Analytics and Intelligent Data-Driven Systems*, https://doi.org/10.1007/978-981-92-2623-8_8

1 Introduction

The expanding skills gap, increasing job flexibility, evolving organizational priorities, and the rise of Artificial Intelligence (AI) are transforming businesses, markets, and the global talent landscape. The creative ideas, problem-solving capacity, and innovative solutions brought in by employees will help any organization to stay competitive and adapt to market changes (Muneer et al., 2024). Employees shape the organizational culture, foster collaboration, and create a positive environment in an organization that in turn enhances the efficiency of the company. Hence, the workforce decisions made today will determine whether any organization thrives or merely survives in the future. A bad hire may lead to frequent turnover, increasing the training and recruitment costs. Hence, to be successful, organizations need "right people with right skill" at the right time to meet the organizational goals (Elugbaju et al., 2024). At this juncture, talent hunting and proper talent acquisition have become a matter of high relevance. Implementation of AI in talent acquisition initiatives is gaining much popularity these days. The practice of employing digital tools, platforms, and technologies to recruit, evaluate, and employ people is known as "digital talent acquisition." In the swiftly changing environment of contemporary organizations, the function of artificial intelligence and digital technologies has become progressively essential for the recruitment process. The implementation of these sophisticated tools has transformed how organizations identify, attract, and assess prospective candidates, resulting in enhanced efficiency, diminished bias, and superior decision-making (Johnson et al., 2021).

AI-driven software and algorithms have facilitated the automation of the preliminary phases of recruitment, including resume evaluation and candidate identification, thereby permitting recruiters to concentrate on the more strategic and interpersonal dimensions of talent acquisition (Rehman et al., 2021). Furthermore, to improve the experience and engagement of candidates in a recruitment process, companies are making use of chatbots and virtual assistants. All the inquiries by the candidate will be given personalized and prompt responses by these AI models. When compared to robotic process automation, AI-based technologies are more powerful. By evaluating a resume, AI models can make their own decisions regarding the tests and assessments to be done and also regarding the selection strategy (Agarwal & Sinha, 2024).

To expedite the hiring process, it makes use of data analytics, Applicant Tracking Systems (ATS), social media, job portals, and AI-driven recruitment technologies. Increased reach, better efficiency, better applicant experience, and data-driven and less biased decision-making are some of the main advantages. To maximize hiring, businesses employ video interviews, chatbots driven by AI, and digital marketing techniques to attract talent.

This study adds to the existing literature by giving an extensive viewpoint on how technology-enabled recruitment strategies are making a change in organization's effectiveness. By analysing how AI, automation, and analytics are changing the phase of talent acquisition from employee branding, sourcing of candidates, to selection and retention, this research aims to strengthen the existing work by providing a comprehensive literature on technology adoption, its effectiveness, and challenges in human resource strategies.

The unique contribution of this study lies in developing a conceptual roadmap for adaptive talent acquisition that takes into account the VUCA (Volatile, Uncertain, Complex, and Ambiguous) world and the dynamics of the Gig Economy. The study's primary objective is to explore how talent acquisition methods are transitioning from conventional, human-driven systems to technology-enabled human strategies that provide a flexible system (Al-Alawi et al., 2023). Secondly, the study also aims to understand how automation, data analytics, and AI have impacted the hiring processes with a due focus on Employee Branding and Employee Value Propositions in attracting and retaining the talent. Also, the aim is to propose a conceptual framework, which is future-centric and is more focused on satisfying the demands of a dynamic, competitive, and agile environment while strategically accounting for various constraints and challenges.

1.1 Role of Enhancing Candidate Experience, Employer Branding, and Employee Value Proposition

Job seekers now expect a seamless experience through interactions with recruiters and technology. Candidates seek user-friendly career sites that allow them to quickly create profiles, search for jobs, and apply effortlessly. Hence, many organizations have upgraded their career sites to streamline applications, personalized interactions, and actively solicit candidate feedback to enhance the experience further. This experience has become one of the enablers for building an authentic employer brand. To attract top talent, companies need a strong employer brand that resonates with the expectations of potential candidates. A well-defined Employee Value Proposition is key to standing out in a competitive market.

Yu Trevor et al.'s (2022) study highlights the relationship between competitive advantage and employer branding. The research focuses on the effects of the macro environment, employer branding strategies and models, the implementation process of employer branding, and its nuances. Kumar (2019) listed five ways in which hiring new employees can support strategic corporate goals. They contend that hiring staff members who share company values is essential. Second, although creating human capital tailored to a particular company has been viewed as a challenging endeavour, proper talent acquisition strategies may indicate how particular types of human capital might thrive in particular organizational environments. Third, by attracting external talent to work alongside internal talent and guaranteeing a continuous renewal of strategic perspectives (Alkhawaldeh et al., 2024). Fourth, by anticipating technology upheavals, the talent acquisition department can find individuals with high learning agility who can assist organizations in navigating these changes. The fifth strategy talks about being cognizant of the business cycles of expansion.

The relationship between employee engagement and HR procedures was found to be mediated by corporate branding which was also propagated by Mathew (2015) in the study on talent management practices in some selected organizations in India. In this study, Verma et al. (2024) experimentally assessed employer branding and the application intentions of prospective employees. As per this study, factors like organization culture, job security, job role, compensation package, and international exposure are the most important factors candidates look into when they apply for a job.

1.2 Talent Acquisition Practices in Organizations

There is a shift from traditional role-based hiring to a skill-focused talent acquisition process, due to changes in business operating models and business environment. Hence, new age practices for skill assessment through various tools and techniques are gaining prominence. Besides this, a new model of acquisition has emerged over the last 4–5 years. Mathew (2015), in the study, gained insights from various professionals to understand their talent acquisition strategies. Using a sample of 30 companies selected based on their global reach, distinctive brand, and established presence, this study has attempted to examine talent management procedures and practices. The result shows that most banks use employer branding as a significant tool in talent acquisition. Several Human Resource (HR) strategies, like building internally, sharing global resources, and flexible work strategies, are also discussed in detail. Anbumathi and Sivasubramanian (2016) in this study examine and talk about Knowledge Management (KM) practices and Talent Management (TM) initiatives using a general review technique. A theoretical model that combines KM and TM techniques has also been put forth. It concludes by developing a research framework on the association between KM and TM practices.

Pocztowski and Pauli (2022) analysed the talent acquisition practices in small and medium enterprises. The analyses aim to ascertain the degree to which talent acquisition practices are influenced by contextual factors, with special focus on small and medium firms. Analysis output shows that there is a high level of correlation between two factors, context and TM, but due to the variety of enterprises used in this study and also due to the changing perspectives of managerial and ownership persons on Human Resource Management (HRM), no consistent patterns are recognized.

1.3 AI in Talent Acquisition

Johnson et al. (2021) in their research paper provide insights regarding how Electronic HRM (eHRM) and AI can be used in the hospitality sector. The study discusses the use of technology in HRM, AI-based recruitment, and selection process and substantiates that AI and eHRM have the power to revolutionize hiring and selection processes across various sectors. They also conclude by mentioning steps to be taken to guarantee that the decisions and insights obtained are well-received by staff members and result in improved organizational and employee results. Rehman et al. (2021) present empirical evidence regarding the factors and outcomes associated with the utilization of social media technology in recruitment. This study utilized a sample size of 589. This study employs tools such as multiple correlation and regression. The output shows that the incorporation of social media recruiting technology into the recruitment model, encompassing all elements and outcomes, influences variables such as recruitment duration, expenses, and quality.

AI supplants monotonous tasks typically performed by human recruiters (Upadhyay & Khandelwal, 2018) and assists HR managers and executives in efficiently attracting, retaining, and motivating skilled HR personnel, thereby benefiting the company's growth (Raviprolu, 2017). It can reduce a company's expenses by enhancing the efficiency of the hiring process, resulting in decreased personnel costs and employee turnover (Buckley et al., 2004). The next advantage pertains to the application of diversity and cultural fit, as unconscious human bias is diminished, facilitating fair assessment. AI can aid in identifying the most qualified candidates by removing unconscious bias from manual selection and evaluation processes, excluding variables such as names, educational institutions, gender, age, and race (Upadhyay & Khandelwal, 2018).

1.4 Challenges in Digital Talent Management

According to Kane and Animesh (2020), the term "digital age" refers to a period of time marked by the widespread use of digital technologies in a variety of spheres of human existence, such as communication, business, education, and entertainment. There is a high degree of change in the way people communicate, do official work, and access information, especially after the implementation of virtual tools like online social networks, smartphones, and cloud-based computing. In addition to this, there is a greater change in the strategies followed by companies to handle their workforce in this technological era. Due to the substantial use of AI technology, strategies of talent searching, selection, growth, and retention have evolved considerably. Because of this, now companies can reach out to larger groups of people and can also communicate with them easily. On-demand learning experiences can be provided with the help of digital tools.

Ladika (2022) has mentioned her view in a feature published by the Society for Human Resource Management that burnout has become a severe concern for employees, especially after COVID-19 pandemic. According to a survey conducted among 726 HR employees in seven different countries, 42% of the respondents are struggling with too many projects and responsibilities (Kumar, 2019). This article also highlights the need for a change in the work environment and strategies, so that the employees can be retained for a longer period. But a major challenge in retaining talent is the need for a strong employer brand. A report published by McKinsey &

Co (2018) revealed that diversity is an important criterion which provides a competitive edge to organizations. Diverse workforce is preferred over companies with less diversity. Hence, inclusive strategies are essential for companies to include when they develop their talent acquisition models. To access a broader candidate pool and establish initiatives that promote diversity and inclusion, companies must leverage digital technologies. An article by Cross and Braswell (2019) explains how data-driven models are useful in hiring more diverse talent. According to them, improving accessibility, awareness, and a system of accountability are driving forces of the digital era, which help in reducing the challenges of talent acquisition.

2 Talent Acquisition Models and Their Application

Talent acquisition models will compile all the resources like technology, people, and the process together for finding out the best candidate, to assess them, to conduct an interview, and also hire them. A good model of talent acquisition includes all these components. Automated platforms are used, which assess the whole process of hiring. It accelerates the speed of the hiring cycle and also helps in connecting with the right candidate. Talent acquisition involves the following steps, which are discussed in the flowchart (Fig. 1).

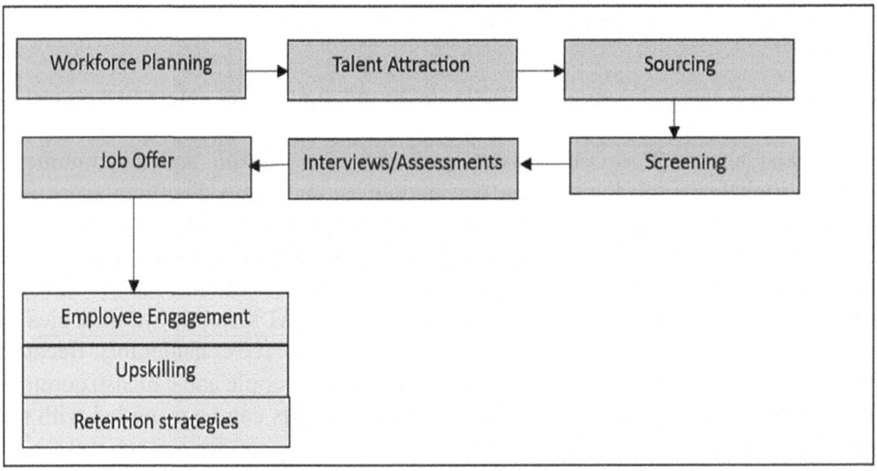

Fig. 1 Talent acquisition process. *Source* Compiled by the researchers

2.1 Structural Approaches to Talent Acquisition

Figure 1 exhibits the process of talent acquisition. The process does not end with a job offer. Companies are also focusing on employee engagement, upskilling employees, and retention strategies. Even though there is no unique model of talent hunting and acquisition, the following are some popular models followed:

2.1.1 In-House Recruitment

It applies to large organizations with regular needs of recruitment. They will have an internal HR team to manage the whole process. The advantage is that the internal team will have a better understanding of the culture and needs of the organization. It is also one of the cost-effective methods.

2.1.2 Recruitment Process Outsourcing (RPO)

This is a mechanism in which the whole or part of the recruitment process is outsourced. A third-party provider will take the responsibility of recruitment and thereby increase the expertise in this area. It can be enterprise RPO, hybrid RPO, and project-based RPO. Enterprise RPO is a large-scale, fully integrated mechanism which takes the responsibility of the recruitment process as a whole. It especially focuses on large organizations and high-volume recruitment. In a hybrid RPO, some aspects of recruitment may be outsourced, but control remains with the internal team. In project-based RPO, the company will have access to some recruiters and can get advice regarding the changing strategies in business and how to train the staff.

2.1.3 Contingent Workforce Model

Here, the members are hired on a temporary, contract, or freelance mode. Normally, there will be a labour pool from which the candidates are selected. It is applicable for seasonal hiring and project-based workforce.

2.1.4 Head Hunting Model

Instead of going for job postings, in this method, a targeted search is done to find the apt person suitable for a specific post in the organization. This is a more difficult process, as it involves targeting a specific individual and is primarily focused on executive-level positions.

2.2 Digital Talent Acquisition in the Global Arena

In the global scenario, especially in the Asia Pacific region, companies are focusing on strategies like understanding the local market, employee branding, and leveraging technology for recruitment. AI-driven applications are widely used for scrutinizing the numerous candidate profiles they receive. Google has a dedicated in-house talent acquisition team that leverages AI and data analytics to streamline the hiring process. They use an AI-driven Applicant tracking system to assess resumes and predict the best fit for roles. In a global scenario, chatbots and gamified assessments are very popular in use. Amazon hires thousands of temporary workers during peak season periods like Black Friday and Christmas. They rely on staffing agencies to fill the short-term vacancies. They follow the contingent workforce model. Headhunting is another popular method followed by global companies. They search for the right candidates to fill their executive-level positions. Companies like Tesla follow this strategy. From a global perspective, the talent acquisition scenario is changing fast with the merging of AI and automation. Some emerging concepts include Tech innovations, Employer Branding, Data analytics in talent acquisition, the concept of remote work, and a focus on soft skills.

2.3 Digital Talent Acquisition: Indian Scenario

Nowadays, the talent hunt trend in India is more focused on finding individuals who can contribute to the organization than just fitting in. Businesses are also considering gender equality, such as elevating female applicants to mid- or senior-level positions. Additionally, businesses have an inclusive policy. The campus recruitment paradigm is India's most well-known talent acquisition strategy. Many IT firms, such as Infosys, Wipro, and TCS, and FMCG companies, such as HUL and ITC, employ management trainees from business schools. In India, the in-house recruitment model is also widely used. Reliance and Tata have strong internal recruitment teams that leverage referrals. Skills mismatch is one of the issues Indian businesses confront when hiring professionals. The gap between what a candidate learns in school and what they must perform after graduation is always caused by a lack of experience and understanding of real-world procedures. For recruiters, this skill mismatch is a significant worry. Upskilling and training the candidate come at a high expense. Including distant workers, Fitting the new pool into the organization's culture has become more difficult due to the rise in remote work following COVID-19.

2.4 Emerging Concepts Gaining Popularity in the Indian Landscape

2.4.1 Employee Value Proposition (EVP)

An employer's beliefs, rewards, recognition, support, and business culture are outlined here to help employees reach their full potential and perform at their maximum level. Under this approach, each employee is viewed as a person, not just a worker. This criterion is used to maintain an organization's culture compared to its rivals. To ensure that the right people join the company and to lower the turnover rate, the ideas of employer branding and EVP are becoming more and more popular in India. Qualtrics.com (2024), on their website, mentions factors contributing to EVP. Health care coverage, paid time off, retirement plans and packages, and child care contributions are some of the novel ideas which can increase employee engagement and loyalty.

2.4.2 Employer Branding

Employer branding is how a business presents itself as an appealing workplace. Because a strong employer brand attracts top talent, lowers costs, and improves employee retention, and it is crucial to talent acquisition. This idea is comparable to that of EVP. For instance, Google's employer branding emphasizes creativity, adaptability, and employee benefits, making it a desirable workplace. The components that businesses provide to develop their branding are (1) Competitive pay and benefits, (2) Opportunities for career advancement, (3) A culture of positivity at work, (4) Work–life harmony, and (5) Inclusion and diversity. Another example of improving employee engagement and employer branding is Advocacy and referrals by Employees. TCS encourages employees to share success stories and work–life experiences on social media platforms to attract new talent. When compared to EVP, Employee branding includes social media presence, company reviews, and employee testimonials. This is more of attracting new talent while EVP focuses on retaining and motivating employees.

2.4.3 Value Addition

Businesses are now searching for individuals who contribute to the organization instead of merely fitting in with the culture. Models for talent acquisition will assist the organization in connecting with individuals with the appropriate skill set and disposition. One indicator that aids in analysing an employee's true contribution is the ETI headcount. The total number of employees on paper is reduced by the number of inactive employees.

3 Evolution of Talent Acquisition: Transforming Models and Strategies

Digital methods have replaced manual processes, newspaper job ads, and in-person interviews in recruitment. Online job portals in the late 1990s and early 2000s changed organizational candidate sourcing by expanding outreach and speeding up application processes (Ployhart, 2006). Technology, labour market dynamics, and business needs have driven talent acquisition models and strategy evolution. Despite this, these initial digital strategies focused on filling vacancies rather than actively recruiting. As workforce needs changed, companies adopted strategic, data-driven hiring methods, including employer branding, candidate experience, and long-term workforce planning (Hmoud & Várallyai, 2019).

Early 2010s witnessed a shift with the introduction of Applicant Tracking Systems (ATS) and the use of social media for recruitment. ATS automated resume evaluation, job applications, and candidate databases to streamline recruitment. LinkedIn, Facebook, and Twitter helped recruiters find talent, engage passive candidates, and boost employer brand visibility. Organizations created engaging online narratives to attract job candidates as "social recruiting" gained popularity. Companies now prioritize workplace culture, employee testimonials, and online reviews as key candidate determinants due to employer branding (Chowdhury et al., 2023). Recruitment marketing began with targeted advertising, content marketing, and personalized engagement to attract top talent.

In the mid-to-late 2010s, AI and big data accelerated recruitment model evolution. Chatbots, predictive analytics, and machine learning algorithms automate candidate screening, match applicants to job descriptions, and reduce hiring biases. Companies started using data analytics to evaluate hiring trends, predict workforce needs, and evaluate recruitment efficacy. Companies are exploring flexible workforce models, contract hiring, and global talent pools due to the gig economy and remote work (Johnson et al., 2021). AI-driven video interviews and virtual assessment centres became popular, especially among multinational corporations, to streamline and standardize hiring processes (Gilch & Sieweke, 2021).

COVID-19 accelerated digital transformation in talent acquisition, forcing companies to adopt remote and virtual hiring. Video interviewing, AI-powered evaluations, and remote onboarding became essential recruitment tools (Bennett & McWhorter, 2021). The concept of "borderless hiring" has grown, allowing companies to hire globally (Olokundun et al., 2021). Diversity, Equity, and Inclusion (DEI) in recruitment strategies became a priority, with AI tools used to reduce biases and promote fair hiring (Al-Alawi et al., 2023). Organizations must rethink their talent strategies and implement digital tools for in-office and remote workers due to the rise of hybrid workforce models (Guerra et al., 2023).

Blockchain, metaverse, and hyper-personalized recruitment will shape talent acquisition. Blockchain technology will improve credential verification, considerably reduce fraud, and increase the recruitment trust among potential applicants. Immersive candidate evaluations and job simulations are investigated using

Virtual Reality (VR) and Augmented Reality (AR). The metaverse opens new doors for virtual recruitment, networking, and AI-enhanced talent marketplaces (Dutta et al., 2023). Predictive analytics and sentiment analysis will improve candidate engagement strategies, personalizing and data-driven recruitment.

As organizations seek greater agility in talent acquisition, balancing automation and human-centred hiring methods will ensure a seamless, ethical, and practical recruitment process. Many organizations in the modern day are adopting a mixed recruitment model, combining the benefits of sophisticated technologies and, at the same time, creating a blend of human touch by facilitating person-to-person interaction during some stages of the recruitment process.

3.1 Gig Economy and Freelance Talent Acquisition

Short-term, project-based, or temporary labour agreements are typical in the modern employment paradigm referred to as the "gig economy," which is also sometimes called the "freelance economy." With more and more qualified professionals choosing contract and freelance employment, the gig economy is changing the nature of work in the future. This dynamic transition leads to both opportunities and challenges for companies while hiring top people. It is highly relevant for HR recruiters and company officials to gain expertise to recruit in a scenario of Gig economy.

In the present world, there is a rise in demand for short-term, project-based employees. People look for a flexible work environment and task-based work profile. This creates challenges for traditional HR procedures. Hence, there is a need to shift from conventional methods to more sophisticated methods to withstand the requirements of the Gig economy. Boocock et al. (2020) mention the problems in creating a competitive advantage over the traditional HR relations while creating a new HR configuration.

As a result, HR management, employment relationships, and employment modalities should all be adaptable to different levels of human capital. This model's description of a conventional employment connection contrasts with the gig economy, which emphasizes a workforce involved in specialized activities and calls for a non-standard relationship (Davis & Sinha, 2021). Based on the utilitarian principle of mutual advantage, non-traditional working connections between companies and gig workers create a transactional and symbiotic relationship, according to the literature on HR management in the gig economy (Lepak & Snell, 1999, 2002).

4 Technology and Automation in Talent Acquisition and Its Impact

With sophisticated technology-enabled systems and software coupled with predictive modelling, human resource requirements can be calculated automatically and candidate profiles could be sourced and screened through AI without any human intervention. Candidates record their video interview through pre-defined questions and AI algorithms analyse the response. Further, the algorithm is designed in such a way that it responds to a set of pre-determined exercises, based on the performance of the candidate, and if the candidate is found to be appropriate, a pre-determined offer is generated following which the candidate is onboarded based on the schedule. This entire process can be performed without any human intervention.

AI has been a constant source of information while automating the recruitment process, which includes activities like sourcing the right candidates, scheduling the interviews, and the initial screening of the resumes. By utilizing these intelligent applications, the hiring process can be more systematic and easily monitored. Machine learning algorithms and data analytics assist organizations in targeting and attracting the right kind of candidates. Further to this, such intelligent systems improve the overall efficiency and accuracy of the hiring process. When considering the potential candidates' viewpoint, the integration of Artificial Intelligence (AI), Machine Learning (ML), big data analytics, and automation has further enhanced the experience of the candidates and makes their application process timely and regulated (Parveen et al., 2025). Human errors and bias can exist in the traditional recruitment process as this involves the psychological, emotional, and physical factors of both the employer and the candidate while recruiting. To a large extent, the automation process can help remove this bias and focus on identifying the most apt candidates for the job (Dutta et al., 2023). Modern organizations exploit the most sophisticated technologies like Natural Language Processing (NLP) and Machine Learning (ML) algorithms to align the candidates according to the job requirements, qualifications, and experience (Wang et al., 2024).

Initial interactions with the candidates can be appropriated through AI chatbots and virtual assistants by clarifying the potential queries of the candidates and assisting them in the application process (Wang et al., 2024). Sophisticated technologies like predictive analytics and AI-driven capabilities can help recruiters to assess the candidate's performance based on their cultural orientation and background (Robert, 2024). Sourcing of candidates through various models like social media platforms, networking sites, application tracking systems Mukul & Saini (2021), and other online sources can be efficiently done with the help of intelligent analysis of historical data on hiring records and sources (Alkhawaldeh et al., 2024; Mukul & Saini, 2021). This can also improve the decision-making capabilities of the HR managers and reduce the physical and human bias that can emerge during the acquisition process (Olokundun et al., 2021; Kurek, 2021). Data analytics can also contribute to understanding workforce trends and patterns and evaluating the suitability and appropriateness of the potential candidates (Agarwal & Sinha, 2024). Utilizing qualitative research techniques like sentiment analysis helps to evaluate the engagement of the candidates, ascertain their personality traits and understand their interests which further contributes to deciding whether the candidate is compatible with the organization's culture. Job advertisements can be creatively made and compiled by using data-driven insights from the markets to create unique and crafted call for applications and job opening brochures and posters that can further lure and attract the right kind of candidates.

The influence of recent advancements in blockchain technology, Virtual Reality (VR), and the metaverse will decide the future of talent acquisition as many organizations are increasingly adopting digital transformation. Increasingly, blockchain-based credential verification systems are being used to more securely and authentically verify applicants' degrees, professional backgrounds, and certificates. This decentralized method helps to reduce the occurrences of resume fraud and speeds up the hiring process by avoiding the necessity for comprehensive background checks. Additionally, research is being done on immersive candidate evaluations using Virtual Reality (VR) and Augmented Reality (AR), which would allow job seekers to participate in realistic job simulations prior to making career decisions (Touriano et al., 2023). The introduction of virtual hiring events and automated and AI-powered talent marketplaces in the metaverse points towards a shift to a highly digital and interactive recruitment framework (Koivunen, 2024). Even though these developments have a lot of promise, businesses need to strike a balance between automation and human engagement so that technology complements rather than replaces the relational and strategic aspects of talent acquisition (Praba et al., 2025).

Automating talent acquisition processes has resulted in notable enhancements in efficiency, especially in high-volume recruitment situations. Robotic Process Automation (RPA) is employed more often to take care of recruitment activities, which are undertaken on a repeat and more regular basis, which include resume screening, background investigations, and compliance validation (Yu et al., 2022). These automation technologies facilitate uniformity in recruitment procedures and also make HR teams concentrate and focus more on strategic decision-making of talent management, including relationship cultivation and employer branding (Johnson et al., 2021). By analysing speech patterns, nonverbal cues, and emotional intelligence, video interviewing platforms with AI-powered facial recognition and audio analysis capabilities enhance the assessment process (Almeida et al., 2025). On the one hand, these technologies provide greater efficiency, accuracy, and objectivity in the whole process of talent management, but ethical issues related to privacy, data security, and possible algorithmic biases. On the other hand, they cannot be overlooked and are vital considerations for organizations implementing AI-driven recruitment solutions (Lyons, 2022).

The incorporation of Artificial Intelligence into the talent acquisition process is not just limited to the assembling of data about applicants but also envisages the choosing of the right kind of candidate who is the most appropriate for the position (Vedapradha et al., 2023). Various studies have been undertaken to assess the acceptance process for technological applications by the end users, and the most

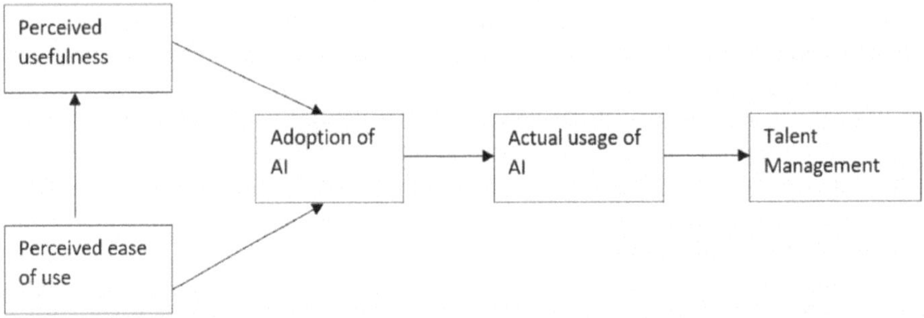

Fig. 2 Intelligent talent acquisition and Technology Acceptance Model. *Source* Vedapradha et al. (2023)

widely accepted conceptual model is the Technology Acceptance Model (TAM) which was developed by Davis (1989). This model consists of two major factors affecting the acceptance level of consumers, which are the perceived usefulness and the perceived ease of use of the technology. As illustrated in Fig. 2, this theory propagates that the potential user's intention to adopt the technology and its actual usage will depend on their perception of the usefulness and the user-friendliness of the application.

Vedapradha et al. (2023) have substantiated through their research that talent acquisition and HR analytics are the most significant components of generating a competitive edge among organizations. Artificial intelligence can be used by businesses to accelerate the levels of investments in integrating big data analytics and data-driven insights, which in turn contribute to better and more efficient decision-making. By making use of the right kind of technology applications in talent acquisition and management, organizations can enhance their capabilities to a much higher level (Fig. 2). Modern-day organizations are largely impacted by their adoption and implementation of technological applications in managing HR data and carefully utilizing this data to scrutinize the applications.

5 Methodology

This study adopts PRISMA-based systematic review of existing literature to examine the evolution and adoption of talent acquisition models in the dynamic business environment. The Preferred Reporting Items for Systematic reviews and Meta Analysis (PRISMA) model was adopted for the study considering the benefits of using a structured framework to conduct and report literature related to the study in a transparent and replicable manner. The methodological guidelines of the PRISMA model help to identify, screen, evaluate, and include all the relevant literature, giving unbiased and transparent results (Moher et al., 2010). Research credibility and reduction of bias are considered the major attributes of this model, even though the process is time-consuming and generates large data volumes, adding to the complexities of the process (Panic et al., 2013).

5.1 Synthesis of Literature

The methodology is a synthesis of secondary data from peer-reviewed journals, books, case studies, and industry reports. The search also focused on previous studies addressing talent acquisitions, digital recruitment technologies, and emerging recruitment practices. Key search terms were talent acquisition models, machine learning in the recruitment process, hiring practices in Gig economy, and artificial intelligence models.

5.2 Screening and Selection Criteria

Titles and abstracts of the initial pool of studies were reviewed to ensure the relevance of the research. Studies not directly related to talent acquisition technologies were removed during this stage of screening. Full text articles were assessed in detail and were included if they a) discussed the integration of technology in the recruitment process, b) provided empirical or conceptual insights into the talent acquisition models, and c) addressed current trends, challenges, or best practices in digital hiring.

5.3 Comparative Analysis

Relevant data from the selected studies were extracted and analysed. A comparative analysis was conducted to examine the integration of various technologies such as Artificial Intelligence, Machine Learning, and Digital Automation tools, to acquire potential talents from across the globe. The study also analysed the empirical findings developed from past research to highlight the current trends, challenges, and best practices followed in the industry.

5.4 Thematic and Content Analysis

Content analysis was performed to identify the major themes emerging in the research. Key themes include digitization, potential candidate experience, and the impact of the gig economy on talent acquisition strategies. The study has also used theoretical frameworks such as the Technology Acceptance Model (TAM) to interpret the adoption of technological models in the recruitment process.

6 Results and Discussions

Based on the systematic screening of literature using the PRISMA framework, recurring concepts related to talent acquisition practices were identified and analysed. As a result, key themes associated with digital transformation, emerging models, and technological applications were identified. The findings are summarized in Tables 1 and 2.

Table 1 indicates the key emerging themes on digital talent acquisition from the pertinent literature on the fields of Human Resource Management (HRM). The significant theme being the involvement of Artificial Intelligence (AI) and automation in recruiting processes across the various firms, improving efficiency and facilitating seamless, transparent selection methods. Employer and EVP are another key theme that highlights the emerging prospects of digital talent acquisition and retention of high-quality labour force in the competitive market. The emergence of data-driven decision-making tools, application tracking systems, and analytics-based recruitment has resulted in a highly structured and standardized human resource management strategy.

There is a clear indication from the literature on the transition of traditional hiring and recruitment strategies to the automated and technology-assisted process in

Table 1 Thematic analysis of talent acquisition transformation

Theme	Key concepts	Related studies	Implications
AI driven recruitment	Automated screening, predictive analytics, resume parsing	Johnson et al. (2021) and Rehman et al. (2021)	Improved efficiency and speed, data driven decision making, improved candidate matching
Employer branding	Organisational reputation, EVP, digital branding	Yu Trevor et al. (2022) and Verma et al. (2024)	Attract high quality talents, competitive advantage in talent markets
Data driven hiring	ATS systems, analytics-based decision making	Buckley et al. (2004) and Upadhyay and Khandelwal (2018)	Improves recruitment accuracy, increase transparency, strategic workforce planning
Candidate experience	AI chatbots, personalized communication	Johnson et al. (2021)	Improved candidate engagement, enhances employer brand perception, faster response
Gig economy integration	Freelancers, contingent workforce	Kumar (2019)	Increase work force flexibility, more global access, faster project execution

Source Compiled by authors

Table 2 Talent acquisition models in comparison with traditional model

Dimension	Traditional model	Technology enabled model	Adaptive talent acquisition model	Relevant literature
Recruitment process	Manual screening	Applicant tracking system and AI tools	Integrate data analytics, predictive analytics, automation	Upadhyay and Khandelwal (2018)
Candidate sourcing	Job advertisements	Social media recruiting platforms, automated sourcing tools	Multi channel intelligent sourcing, AI algorithm, gig platform integration	Wang et al. (2024)
Decision making	Human judgement	AI supported decision making, Data dashboards	Data driven predictive analytics, real time labor market analytics	Yu et al. (2022)
Candidate engagement	Email communication	Chatbots and virtual assistant, CRM platforms	Personalized digital engagement	Almeida et al. (2025)

Source Compiled by authors

recent years. Table 2 summarizes the developments and trends in the human resource policies and strategies over the past decade. Traditional recruitment models largely relied on manual screenings of advertisements and communications which required a high degree of human intervention. Over the years, this has shifted to integrated AI tools, automated screening, Intelligent sourcing, integration of Gig economy, predictive analytics to forecast labour force requirements, and personalized employee engagement.

6.1 Technology Applications in Talent Acquisition

Content analysis of the previous studies also revealed that there is an increasing integration of digital technologies in talent acquisition practices. Several studies emphasized the application of AI and Application Tracking System in automating the recruitment process which enables the predictive hiring mechanism. This will result in systematically tracking and evaluating applicant information (Davis & Sinha, 2021; Elugbaju et al., 2024). AI-powered chatbots also emerged as a prominent tool for candidate engagement and communication (Koivunen, 2024). Another technology discussed in a few studies is the use of Virtual Reality (VR) technologies used for job previews and candidate assessments, allowing organizations to provide realistic job insights (Lyons, 2022). Application of data analytics is also widely discussed in papers, as a tool for providing strategic workforce planning (Wang et al., 2024). All findings show a clear shift towards technology-driven recruitment practices aimed at improving efficiency, decision quality, and candidate experience.

6.2 Challenges in Implementing Talent Acquisition Models

A few studies also discussed the challenges in implementing talent acquisition models. Several studies indicated the concept of VUCA (Volatility, Uncertainty, Complexity, and Ambiguity) environment in shaping the contemporary talent management and recruitment process. Several studies indicate that the concept gained prominence after the global financial crisis and has been widely used to describe the turbulent and rapidly changing business landscape (Bennett & Lemoine, 2014). The literature further suggests that rapid technological advancements, particularly in artificial intelligence and automation, are significantly transforming organizational cultures and human resource management practices (Friedman, 2005). These developments demonstrate that HR systems are increasingly dynamic and influenced by continuously evolving internal and external environmental conditions. Furthermore, the synthesis of prior studies suggests that leadership development and talent management practices must be redesigned to align with the demands of the VUCA environment. Organizations increasingly require leaders who are capable of embracing uncertainty, responding to change, and demonstrating critical thinking and strategic adaptability (Venkatesh, 2016). The VUCA framework is therefore identified as a useful model for HR professionals to evaluate the internal and external factors affecting organizations and to design leadership development initiatives accordingly. In this context, HR departments play a crucial role in preparing future leaders who can navigate complex and competitive environments (Mukherjee, 2020).

7 Road Ahead: Future of Hiring Strategies

The initial step towards developing a strong workforce is to align human resources of an organization with its functional and strategic goals. This alignment will allow the Human Resource (HR) department to make informed decisions about the workforce needs, such as the skills that the company requires and the number of employees that have to be recruited and the time when the recruitment should take place. Strategic workforce planning is very critical in this process since it determines the current trends of the workforce and predicts future needs of talent. In this manner, the organizations will be able to build a suitable hiring strategy which is highly integrated with the other organizational functions.

Other components that are included in effective workforce planning are workforce analytics, succession planning, talent management, and performance evaluation systems. These factors enable the organizations to have systematic checks on the performance of employees, leadership pipeline, and continuity of the key positions. Also, the insight tools like snapshot surveys or quick check-ins may be used to give timely data regarding employee engagement and the work environment. Analytical tools that are driven by data also augment decision-making because they help organizations to determine the probability of employee attrition of a high-performing employee and develop adequate retention measures.

In the modern business world, data has emerged as a pivotal resource in nearly all functional sectors such as talent acquisition. To be more strategic and efficient in recruitment, organizations tend to rely more on the data-driven approach. In the case of large organizations, in particular, even simple information, such as being able to keep track of the composition of the workforce and role allocation, may become complicated and hard to handle. Consequently, technological systems, which can categorize and analyse workforce information, are necessary.

Increasingly, modern talent acquisition systems are built to measure not only technical or hard skills but also soft skills which are becoming relevant in a world where artificial intelligence (AI) and technologies are changing the demands of work in a radically short time. With the automation of some repetitive technical competencies, new skills like adaptability, interpersonal, teamwork, and problem-solving competencies are gaining importance. It is therefore up to the organization to review its current workforce capacity and formulate workforce initiatives that will integrate both hard and soft skills evaluation.

The shift towards sophisticated practices of acquiring talent as opposed to traditional recruitment is also being transformed as a result of technological advancements. The introduction of AI and Robotic Process Automation (RPA) has made most of the recruitment-related activities much easier. Administrative processes that are routine (including report generation, basic application screening, and confirmation mails) can now be automated. This automation saves on man-hours of work and enables the HR professionals to spend more time on strategic planning, workforce analytics, and making decisions that are based on data.

Over the past several years, the area of talent acquisition has experienced significant changes, which have been contributed to by both technological change and the changing expectations of employees. These transformations are bound to continue in the long run. This means that companies need to be strategic and data-driven when it comes to recruitment so as to keep up with the pace of attracting and retaining the best employees. Lack of adaptation to these changes can make the organization be left out in the current competitive talent market.

8 Conclusion

In conclusion, the present study provides a comprehensive synthesis of existing literature on talent acquisition models in the context of dynamic and technology-driven business environments. Through the integration of PRISMA-based literature review, the study represents key trends revolving around the recruitment processes that can further enhance efficiency and decision-making. Findings of thematic and content analysis highlight a significant transition from the traditional recruitment process to technology-enabled models supported by artificial intelligence, data analytics, digital automation, and advanced recruitment platforms. With organizations shifting their focus from conventional recruitment methods to highly sophisticated and technology-driven, data-driven futuristic models, this research confers a clear overview of the current trends in the field of HRM. The study contributes to the existing knowledge sharing system on technology-enabled digital talent acquisition, providing a conceptual base for further empirical investigations in this evolving area of research.

References

Agarwal, P., & Sinha, J. (2024). Digitizing talent acquisition: A comprehensive analysis of E recruitment practices and trends. *Anvesak, 54*(1), 32–37.

Al-Alawi, A. I., Messaadia, M., Mehrotra, A., Sanosi, S. K., Elias, H., & Althawadi, A. H. (2023). Digital transformation adoption in human resources management during COVID-19. *Arab Gulf Journal of Scientific Research, 41*(4), 446–461. https://doi.org/10.1108/AGJSR-05-2022-0069

Alkhawaldeh, M. M. K., Ali, A. J. B., Mohammad, A. A. S., Shatnawi, T. M., Al-Momani, A. A., Alzoubi, A., et al. (2024). The impact of artificial intelligence on talent acquisition strategies in Jordanian technology startups. In *Business analytical capabilities and artificial intelligence-enabled analytics: Applications and challenges in the digital era* (Vol. 1, pp. 309–323). Springer Nature.

Almeida, F., Junça Silva, A., Lopes, S. L., & Braz, I. (2025). Understanding recruiters' acceptance of artificial intelligence: Insights from the Technology Acceptance Model. *Applied Sciences, 15*(2), 746. https://doi.org/10.3390/app15020746

Anbumathi, R., & Sivasubramanian, R. C. (2016). Amalgamation of talent and knowledge management practices: A conceptual research model. *Vision: The Journal of Business Perspective, 20*(4), 278–289. https://doi.org/10.1177/0972262916668711

Bennett, E. E., & McWhorter, R. R. (2021). Virtual HRD's role in crisis and the post Covid-19 professional lifeworld: Accelerating skills for digital transformation. *Advances in Developing Human Resources, 23*(1), 5–25. https://doi.org/10.1177/1523422320973288

Bennett, N., & Lemoine, J. (2014). What VUCA really means for you. *Harvard Business Review, 92*(1/2). Retrieved October, 2024, from https://papers.ssrn.com/sol3/papers.cfm?abstract_id=2389563

Boocock, A., Page-Tickell, R., & Yerby, E. (2020). The dis-evolution of strategic HRM in the gig economy from talent management to supply chain manager. In *Conflict and shifting boundaries in the gig economy: An interdisciplinary analysis*. Emerald Publishing. https://doi.org/10.1108/978-1-83867-603-220201008

Buckley, P., Minette, K., Joy, D., & Michaels, J. (2004). The use of an automated employment recruiting and screening system for temporary professional employees: A case study. *Human Resource Management, 43*(2), 233–241. https://doi.org/10.1002/hrm.20017

Chowdhury, S., Dey, P., Joel-Edgar, S., Bhattacharya, S., Rodriguez-Espindola, O., Abadie, A., & Truong, L. (2023). Unlocking the value of artificial intelligence in human resource management through AI capability framework. *Human Resource Management Review, 33*(1), 100899. https://doi.org/10.1016/j.hrmr.2022.100899

Davis, F. D., Bagozzi, R. P., & Warshaw, P. R. (1989). User acceptance of computer technology: A comparison of two theoretical models. *Management Science, 35*, 982–1003. https://doi.org/10.1287/mnsc.35.8.982

Davis, G. F., & Sinha, A. (2021). Varieties of Uberization: How technology and institutions change the organization (s) of late capitalism. *Organization Theory, 2*(1). https://doi.org/10.1177/2631787721995198

Dutta, D., Srivastava, Y., & Singh, E. (2023). Metaverse in the tourism sector for talent management: A technology in practice lens. *Information Technology & Tourism, 25*(3), 331–365. https://doi.org/10.1007/s40558-023-00258-9

Elugbaju, W. K., Okeke, N. I., & Alabi, O. A. (2024). Human resource analytics as a strategic tool for workforce planning and succession management. *International Journal of Engineering Research and Development, 20*(11), 744–756.

Friedman, T. L. (2005). *The world is flat: A brief history of the twenty-first century*. Farrar, Straus, and Giroux. Retrieved October, 2024 from https://media-1.carnegiecouncil.org/import/studio/The_World_Is_Flat.pdf

Hmoud, B. I. F., & Várallyai, L. (2019). Will artificial intelligence take over human resources recruitment and selection? *Network Intelligence Studies, Romanian Foundation for Business Intelligence, Editorial Department, 13*, 21–30. https://doi.org/10.1016/j.hrmr.2022.100899

Johnson, R. D., Stone, D. L., & Lukaszewski, K. M. (2021). The benefits of eHRM and AI for talent acquisition. *Journal of Tourism Futures, 7*(1), 40–52. https://doi.org/10.1108/JTF-02-2020-0013

Kane, G. C., & Animesh, A. (2020). The rise of digital-age organizations: A conceptual foundation. *MIS Quarterly, 44*(2), 449–473.

Koivunen, S. (2024). *Digitalization of talent acquisition: Understanding human resource management professionals' experiences and practices*. Computing Sciences, TAUCHI Research Center. ISBN (Electronic) 978-952-03-3271-6.

Kumar, A. P. S. (2019). Aligning talent acquisition practices to strategic business objectives. *NHRD Network Journal, 12*(3), 261–269. https://doi.org/10.1177/2631454119888092

Kurek, D. (2021). Use of modern IT solutions in the HRM activities: Process automation and digital employer branding. *European Research Studies Journal, 24*(1), 152–170. https://doi.org/10.35808/ersj/2035

Ladika, S. (2022, March 14). *Burnout is a problem for HR professionals*. Society for Human Resource Management (SHRM). https://www.shrm.org/in/topics-tools/news/hr-magazine/burnout-problem-hr-professionals

Lepak, D. P., & Snell, S. A. (1999). The human resource architecture: Toward a theory of human capital allocation and development. *Academy of Management Review, 24*(1), 31–48. https://doi.org/10.1177/014920630202800403

Lepak, D. P., & Snell, S. A. (2002). Examining the human resource architecture: The relationships among human capital, employment, and human resource configurations. *Journal of Management, 28*(4), 517–543. https://doi.org/10.1177/014920630202800403

Lyons, N. (2022). Talent acquisition and management, immersive work environments, and machine vision algorithms in the virtual economy of the metaverse. *Psychosociological Issues in Human Resource Management, 10*(1), 121–134. https://doi.org/10.22381/pihrm10120229

Mathew, A. (2015). Talent management practices in select organizations in India. *Global Business Review, 16*(1), 137–150. https://doi.org/10.1177/0972150914553527

McKinsey Global Institute. (2018). *Jobs lost, jobs gained: What the future of work will mean for jobs, skills, and wages*. https://www.mckinsey.com/featured-insights/future-of-work/jobs-lost-jobsgained-what-the-future-of-work-will-mean-for-jobs-skills-and-wages.

Moher, D., Liberati, A., Tetzlaff, J., & Altman, D. G. (2010). Preferred reporting items for systematic reviews and meta-analyses: The PRISMA statement. *International Journal of Surgery, 8*(5), 336–341. https://doi.org/10.1016/j.ijsu.2010.02.007

Mukherjee, A. S. (2020). *Leading in the digital world: How to foster creativity, collaboration, and inclusivity*. MIT Press.

Mukul, K., & Saini, G. K. (2021). Talent acquisition in startups in India: The role of social capital. *Journal of Entrepreneurship in Emerging Economies, 13*(5), 1235–1261. https://doi.org/10.1108/JEEE-04-2020-0086

Muneer, S., Singh, A., Choudhary, M. H., & Alshammari, A. S. (2024). The mediating role of psychological empowerment on the relationship between digital transformation, innovative work behavior, and organizational financial performance. *Behavioral Sciences, 15*(1), 5. https://doi.org/10.3390/bs15010005

Olokundun, M., Ibidunni, S., Ogbari, M., Falola, H., & Salau, O. (2021). COVID-19 pandemic and antecedents for digital transformation in the workplace: A conceptual framework. *Open Access Macedonian Journal of Medical Sciences, 9*(F), 41–46. https://doi.org/10.3889/oamjms.2021.4952

Panic, N., Leoncini, E., De Belvis, G., Ricciardi, W., & Boccia, S. (2013). Evaluation of the endorsement of the preferred reporting items for systematic reviews and meta-analysis (PRISMA) statement on the quality of published systematic review and meta-analyses. *PLoS One, 8*(12), e83138. https://doi.org/10.1371/journal.pone.0083138

Parveen, M., Suganya, G., Ayisha, M. L. N., & Batt, M. R. P. (2025). The role of artificial intelligence in modern recruitment and selection process. *Cuestiones de Fisioterapia, 54*(2), 630–648. https://doi.org/10.48047/CU.

Ployhart, R. E. (2006). Staffing in the 21st century: New challenges and strategic opportunities. *Journal of Management, 32*, 868–897. https://doi.org/10.1177/0149206306293625

Pocztowski, A., & Pauli, U. (2022). The impact of contextual factors on talent management practices in SMEs. *Human Systems Management, 41*(1), 87–101. https://doi.org/10.3233/HSM-211174

Praba, K., Vijay, B., Abuya, J. O. O., Poreddy, D., Mulakala, S., & Manoharan, S. (2025). Human resource management practices in the age of industrial automation: Impact, adoption, and future workforce. In *Advancements in intelligent process automation* (pp. 79–102). IGI Global. https://doi.org/10.4018/979-8-3693-5380-6.ch004

Qualtrics. (2024). *2024 employee experience trends report*. https://www.qualtrics.com/2024-employee-experience-trends-report/

Raviprolu, A. (2017). Role of artificial intelligence in recruitment. *International Journal of Engineering Technology, 5*(4). https://doi.org/10.56726/IRJMETS65720

Rehman, S., Ullah, A., Naseem, K., Elahi, A. R., & Erum, H. (2021). Talent acquisition and technology: A step towards sustainable development. *Frontiers in Psychology, 13*, 979991. https://doi.org/10.3389/fpsyg.2022.979991

Robert, A. (2024). *Talent acquisition strategies in human resource management* (No. 13119). EasyChair.

Touriano, D., Sutrisno, S., Kuraesin, A. D., Santosa, S., & Ausat, A. M. A. (2023). The role of information technology in Improving the efficiency and effectiveness of talent management processes. *Jurnal Minfo Polgan, 12*(1), 539–548. https://doi.org/10.33395/jmp.v12i1.12454

Upadhyay, A. K., & Khandelwal, K. (2018). Applying artificial intelligence: Implications for recruitment. *Strategic HR Review, 17*(5), 255–258. https://doi.org/10.1108/SHR-07-2018-0051

Vedapradha, R., Hariharan, R., David Winster Praveenraj, D., Sudha, E., & Ashok, J. (2023, January). Talent acquisition-artificial intelligence to manage recruitment. In *E3S web of conferences* (Vol. 376, p. 05001). EDP Sciences. https://doi.org/10.1051/e3sconf/202337605001.

Venkatesh, D. A. N. (2016). Global talent management strategies for organizational success in a VUCA world—A conceptual framework. *UDAAN: The International Journal of Management Research, 4*(02) https://papers.ssrn.com/sol3/papers.cfm?abstract_id=3187145

Verma, S., Kaur, R., & Bhasin, N. K. (2024). Does employer branding shake application intentions of potential employees? *Global Business Review*. https://doi.org/10.1177/09721509231219148

Wang, L., Haque, R., Hakimi, N., & Guanyu, C. (2024). Examining digital talent acquisition and recruitment strategies to enhance economic Sustainable Development Goals (SDGS) of SME's in China's internet industry. *Journal of Lifestyle and SDGs Review, 4*(3), e02306–e02306.

Yu, K. Y. T., Dineen, B. R., Allen, D. G., & Klotz, A. C. (2022). Winning applicants and influencing job seekers: An introduction to the special issue on employer branding and talent acquisition. *Human Resource Management, 61*(5), 515–524. https://doi.org/10.1002/hrm.22140

Yu, K. Y. T., Goh, K. H., & Kawasaki, S. (2022). Investigating electronic word-of-mouth on social media: An eye-tracking approach. *Human Resource Management, 61*(5), 599–616. https://doi.org/10.1002/hrm.22140

K. A. Dhanya is presently serving as an Associate Professor at the School of Business and Management, Christ University, Bangalore. She holds a PhD in Commerce from the University of Calicut, MPhil from Bharathiar University, M. Com from Calicut University, and an MBA from Madras University. She also qualified UGC-NET and JRF in commerce in 2004. To her credit, she has 20 publications in different national and international journals, including FIVE Scopus-indexed. She has been awarded a Minor Research Project sponsored by the UGC, New Delhi, in 2014, and a Seed money project by Christ University in 2026. She has co-authored two books and edited three books. She has been a resource person in 15 seminars and conferences. She has teaching experience of more than 20 years and has held various academic and administrative positions. She served as Head of the Department of Commerce and Management Studies, Amal College of Advanced Studies, Nilambur, Kerala, from 2010 to 2024. She also served as the IQAC coordinator during 2019–2022. She has undertaken a social project in 2019 for revamping tribals in Nilambur Taluk. Due to her contributions in academia and society, she has been awarded "Vanitha Rathna Puraskar" by Jan Shikshan Sansthan Malappuram. She is a fellow member of the Indian Accounting Association and is a life member of the Indian Society for Training and Development (ISTD). Her area of interest covers Financial Management, Capital Market, Corporate Governance, Research Methodology, and Econometrics.

Shikha Malhotra is an expert in Financial Risk Management, Banking, Portfolio Management and Business Valuation. She holds a master's degree in Commerce and is FRM certified by GARP. She has also passed CFA Level Two from CFA (US) and holds a PhD from Jain University, Bangalore. Her research focuses on financial cycles, spillover effects of monetary policies, and risk management in emerging markets. Dr. Malhotra has worked with the State Bank of India for 9.5 years across various domains. She has published papers in ABDC (B and C) journals and Scopus-indexed journals and presented at conferences organized by IIM Bangalore and ICFAI Hyderabad. She has also coached students for competitive exams, including RBI, SEBI, and bank exams. Currently, Dr. Malhotra serves as an Assistant Professor at the business and management, Bangalore Bannerghatta Road Campus.

H. Sandhya is a distinguished academician and researcher currently serving as an Assistant Professor at Christ (Deemed to be University), Bangalore. She holds a PhD in Tourism from the School of Business and Management at Christ University, Bangalore. A testament to her academic excellence, Dr. Sandhya completed her MPhil in Tourism and a Master's in Tourism Administration (MTA) from Christ University, securing the first rank in both programs. Dr. Sandhya brings a unique blend of industry and academic experience. She has worked with Kuoni SOTC, a leading multinational tour operations company in Bangalore, gaining invaluable insights into the tourism sector. She has over seven years of teaching experience, having held positions at various prestigious colleges and institutions in Kerala. She has also contributed significantly to research, serving as a Research Associate for a Major Research Project on "Strategic Intervention of Destination Management Organizations—State of Karnataka". Her research interests span Tourism Marketing, Destination Management, Smart Tourism, and Destination Branding. Dr. Sandhya has made substantial contributions to academic literature, with over 10 publications in Scopus-indexed journals and peer-reviewed books.

Cybersecurity Vulnerability and Organizational Performance: A Systematic Literature Review

Ibrahim Olanrewaju Lawal, Evwiekpamare Fidelis Olori, and Ismaila Yusuf

Abstract The impact of cybersecurity vulnerability on organizational performance is significant in terms of vulnerable sensitive information, disruption of operations, higher costs, and lack of confidence. Effective security measures enhance productivity, resilience, and organizational efficiency in general. This chapter researches the cybersecurity weaknesses and how this weakness affects the organizational efficiency. To examine the associations between cybersecurity threats, such as those of hacktivism, cybercrime, and cyber spying, and the key performance indicators, such as reputational damage, operational effectiveness, and financial stability, a systematic literature review method was used. Data were collected via EBSCOhost, Elsevier, ProQuest, Wiley-20, SpringerLink, and Google Scholar and were reviewed and analysed through Systematic Reviews and Meta-analyses (PRISMA). As technology and information systems expand globally, businesses must stay ahead of the curve by enhancing security measures and proactively resolving authentication and authorization vulnerabilities that could jeopardize confidentiality and availability. Research shows that hacktivism, cybercrime, and cyber espionage harm organizational performance. To combat these risks, companies must create backup systems to protect client and customer information from hostile threats.

Keywords Financial stability · Reputational damage · Cyber espionage · Cybercrime · Operational effectiveness · Hacktivism

I. O. Lawal
Department of Business Management, GBS/University of Suffolk, Birmingham Campus, Birmingham, United Kingdom

I. O. Lawal (✉) · E. F. Olori
Department of Business and Entrepreneurship, GBS, Oxford Brookes University, Oxford, UK
e-mail: ilawal@globalbanking.ac.uk; lawal.olanrewaju@kwasu.edu.ng

E. F. Olori
e-mail: EOlori@globalbanking.ac.uk

I. Yusuf
Department of Business and Entrepreneurship, Kwara State University, Malete, Nigeria
e-mail: ismaila.yusuf@kwasu.edu.ng

M. Dzingirai and S. O. Atiku (eds.), *Strategic Human Resource Analytics and Intelligent Data-Driven Systems*, https://doi.org/10.1007/978-981-92-2623-8_9

1 Introduction

Cybersecurity is a topic explored by numerous authors and experts, and their insights are unique (Rathaus, 2019; Siddique & Rehman, 2017; Sterlin, 2017). Cybersecurity is now a threat to the national security of every nation, even technologically advanced nations like the United States (Ibikunle & Eweniyi, 2020). They state that to achieve global cybersecurity, all countries are required to enact the correct legislation against the misuse of Information and Communication Technology (ICT), including any actions that jeopardize the integrity of critical information infrastructures (Moses-Oke, 2019). Moses-Oke (2019) assert that since threats may occur anywhere globally, the challenges are inherently international and therefore require international assistance, investigative assistance, and substantial and procedural provisions.

2 Background

Technology has become a growth engine, an innovator, and a source of competitiveness in industries (Kamel, 2020). Information and Communication Technology (ICT) growth improved organizational responsiveness as well as the ease of services (Dunmade et al., 2023; Loonam & O'Loughlin, 2021). In the financial sector, the adoption of electronic systems and online payment has boosted efficiency but also left it vulnerable to cyberattacks (Amboko & Wagoki, 2022). The rapid pace of digitalization has brought in vulnerabilities that compromise data availability, confidentiality, and integrity (Kitheka, 2019). As such, information security management has become essential to the survivability of organizations. Cyber threats persist because available technologies remain vulnerable to faults that are not completely removed with software tools available today (Lindskog, 2016). The frequency and nature of cyberattacks keep rising across the globe, which calls for immediate detection and elimination of weaknesses. Despite the institution of various mechanisms, many organizations still experience internal security incidents (Wasilwa, 2019). The attacks come with huge monetary and image losses, especially in emerging economies.

ICT contributes significantly to development worldwide, yet its expanded use exposes more people and systems to cyberattacks (Rosewarne, 2019). With the introduction of information technology in finance, accounting, and governance, there have been intricate issues of security (Ernst & Young, 2018). Cybercrime, such as digital fraud and data leakage, is common, leading to enormous economic losses (Gercke, 2022). In Nigeria, cybercrime has damaged the country's international reputation and chased away foreign investments. The Central Bank's cashless policy and the rapid rise in mobile communications have unintentionally created avenues for cyber fraud (Odunfa, 2021). Nigerian banks reportedly lost billions of naira to cyberattacks from the year 2000, with the estimated yearly loss being over NGN 400 billion (Ali et al., 2024). Aside from financial loss, such incidents destroy confidence in financial institutions and are threats to national security (Ojeka et al., 2017). Other issues which demand high-security models in cyber-physical infrastructure are also challenges (Odhiambo, 2018). Response systems must be effective and technology threat analysis should be made in an orderly manner, and a resilient infrastructure should be adopted. However, most organizations lack sufficient mechanisms to handle cyber risks (James & Thomas, 2022), and existing cybersecurity tools are still inadequate for advanced, networked environments (Whyte, 2022).

Organizational performance, as measured by productivity, timeliness, and efficiency, can be severely affected by cyber incidents (Mathis & Jackson, 2017). Performance outcomes reflect the extent to which employees and systems are contributing towards strategic goals (Elnaga & Imran, 2018; Venkatraman & Ramanujam, 2016). Incidents violate business continuity, corporate reputation, and profitability, thereby eroding stakeholder trust (Lindskog, 2016). Hacktivism has been researched previously and its organizational reputation consequences (Adediwura & Tayo, 2022; Ajila, 2022; Carrapico & Barrinha, 2022a, 2022b; Kitheka, 2023; Wasilwa, 2019). Empirical evidence in developing countries is scarce. The position of unauthorized access in reputational damage in service-based industries in Nigeria-similar settings is still inadequately explained by research (Raggad, 2018). In response to this shortfall, the present research investigates the effect of hacktivism on organizational reputation in Nigerian organizations, whose declining performance and profitability are increasingly linked with cybersecurity infractions. Such breaches have destroyed investor confidence and trust in leading economic institutions.

3 Concept of Cybersecurity

The terms cybersecurity and information security are often used interchangeably, though they are not identical (Ogut et al., 2022; Von Solms & Van Niekerk, 2018). To know their difference, it is necessary to consider both concepts. Information security focuses on protecting information resources such as data, hardware, software, and networks from unauthorized access (Raggad, 2018). The major desire it has is that only authorized individuals or systems can access information required by the organizational objectives. This model goes hand in hand with the business management principles in which the authorization decision is used to promote the strategic goals.

Cybersecurity, on the other hand, extends beyond information security by addressing the prevention and mitigation of organizational damage through control of unauthorized access and reduction of security incidents (Christou, 2016). As defined in ISO/IEC 27000 (2016), data confidentiality, integrity, and availability (the CIA triad) are guaranteed through cybersecurity. All three of these factors are the focus of both cybersecurity and information security. However, some scholars expand this to include authentication and non-repudiation, forming the "Security Star" to reflect modern digital realities (Raggad, 2010; Whitman & Mattord, 2019). Authentication is used to confirm who a user or a system is, and non-repudiation is used to make sure that a sender does not repudiate the information they have sent. In combination, these aspects contribute to the continuity of the business and organizational trust.

The term cybersecurity may, in turn, be seen as the whole process of securing data, systems, and networks against cyberattacks that take place in cyberspace (Jung, 2019). The International Telecommunication Union (ITU, 2018) defines it as a collection of tools, policies, guidelines, and practices that safeguard digital environments, organizations, and user assets Although cybersecurity and information security focus on ensuring confidentiality, integrity, and availability, the former is a wider concept considers social, legal, and technological aspects. It also realizes that digital protection should not only be limited to data but also to individuals and society as well (Safa et al., 2021; Von Solms & Van Niekerk, 2018).

Within recent years, technological changes, including Bring Your Own Device (BYOD), cloud computing, and remote work, have erased the customary organizational boundaries (Chang & Ho, 2019). With the growing networked and decentralization of information systems, cybersecurity offers a framework of a better fit than information security to describe the current-day challenges. Such an expanded view is especially applicable to the developing world, whereby the infrastructural and regulatory variations define the scope of cybersecurity concerns.

Cybersecurity systems in developed nations are sometimes backed by good digital infrastructure, legal frameworks, and powerful institutional capacity. The environments allow more sophisticated detection systems, periodic audits, and enforcement. On the other hand, infrastructural vulnerabilities, lack of technical skills, low awareness, and inconsistency in the enforcement of policies are contributing factors to cybersecurity issues in most developing nations. The opportunities have grown due to the rise of mobile connectivity, digital banking, and e-commerce in Africa, but have also been exposed to threats like identity theft, online fraud, and ransomware (Suleiman & Olayinka, 2021). Such threats use the points of weakness in lax regulatory frameworks, inadequately resourced institutions, and a lack of cybersecurity awareness.

Besides, informal economies, inadequate data protection regulations, and poor usage of encryption technologies define the cybersecurity environment in Africa. Most Small and Medium Enterprises (SMEs) that dominate African economies do not have any structured cybersecurity policies or trained Information Technology experts. In turn, domestic threats are as significant as the external attacks. By contrast, advanced persistent threats, cyber espionage, and state-sponsored attacks are of more importance to developed countries as they have highly digitized environments and rely on critical infrastructure (Carrapico & Barrinha, 2022a).

The growing use of digital platforms in developing countries also exacerbates the problem of jurisdictional overlaps and poor international cooperation. The likes of phishing, financial schemes, and malware attacks are cybercrimes that are notoriously cross-border and most African countries lack extensive cybercrime laws or cross-border mechanisms of enforcement. Though there are some frameworks like the African Union Convention on Cybersecurity and Personal Data Protection (Malabo Convention, 2014), the enforcement is also not even among the member states.

Under such contextual differences, cybersecurity cannot be represented as a universal paradigm. It must be applied in accordance with socio-economic, infrastructural, and cultural realities of the region. In developed countries, the focus is on technological advancement and dependability of the critical national infrastructure; in developing countries, the focus is also on awareness creation, capacity building, and harmonization of regulatory boundaries. Thus, the concept of cybersecurity is relevant in this study because it summarizes both internal and external functions needed to protect organizational and societal assets against the dynamic cyber threats. It does not investigate only the internal security frameworks of organizations but also their partnerships with the government and other private actors that have a role in digital resilience.

To further comprehend cybersecurity, it is necessary to understand its operational domain, cyberspace. The term "cyberspace," first introduced by William Gibson in 1984, refers to a virtual environment where digital information is exchanged (Ahokas et al., 2017). This definition has changed over time to incorporate the technological, human, and communicative aspects developing as facilitators of information sharing (Rantapelkonen & Kantola, 2018). Cybersecurity, in turn, deals with not only technical protection of networks and systems but also with human users who may either be the cause of exposing the systems to threat or may fall prey to an attack themselves (NESA, 2018; Von Solms & Van Niekerk, 2018).

Cybersecurity is a practice that entails protecting digital resources against the threats that undermine confidentiality, integrity, and availability (Boyes et al., 2016). A proper approach to cybersecurity guarantees that personal, corporate, and governmental data is not at risk of unauthorized access or use (Borum et al., 2022; Tucci, 2017). However, the intensity, form, and motivation of cyber threats vary significantly between regions. While cyber espionage, cyberwarfare, and hacktivism dominate the Global North, cybercrime targeting financial systems, mobile networks, and government databases remains the major concern in Africa and other parts of the developing world (Kapto, 2023).

Such differences do not exclude the fact that cybersecurity in any setting is closely connected with human factors. Researchers have found that due to insufficient training, insider carelessness, or laxity in enforcing security policies, a significant amount of security breaches occur (Anderson, 2016; Lindskog, 2016). Another challenge of cybersecurity education and awareness programs that developing countries have undergone is a limited investment; the employees are the weakest link of the security chain. Capacity-building programs are crucial in such settings for long-term resilience (Ajila, 2022).

4 Organizational Performance

When organizations are formed, it is likely that they have a purpose and that they are working towards achieving a desirable future state. Performance measures how well an organization performs in achieving its goals (Stoner et al., 2016). According to the writers, an organization's performance is a gauge of how well it performs its duties. Stated differently, the organization tries to evaluate how well it accomplishes its relevant goals. The ability of a company to use its resources effectively to accomplish outcomes and reach its objectives is known as corporate performance (Wokoma & Iheriohanma, 2021). Every aspect of operations, both inside and outside the organization, is impacted by the idea of performance. Since the results of current decisions are projected into the future, businesses must analyse performance in all areas, including financial, production, marketing, management, and general activities (Oparanma, 2021).

Performance evaluation is a continuous, ongoing process for many businesses. Executives are increasingly using corporate performance dashboards to monitor important performance indicators, such as sales compared to goals, the quantity of backordered products, or the proportion of customer support enquiries that are addressed within predetermined time frames (Bartholomew, 2020). According to the author, dashboards that are implemented for every individual in an organization will assist in monitoring progress towards objectives, identifying performance gaps, and developing creative solutions to get back on track. According to Ojo (2021), empirical research on business performance originates from fields including accounting, finance, strategic management, and industrial economics. Therefore, the focus in each case seems to be determined by the evaluation's aim and who is conducting it. The ability of a business to achieve beneficial results in terms of productivity level and, of course, profitability, personnel turnover rate, and innovativeness is known as organizational performance. The primary variable of interest in this review is organizational performance. It is thought to be a component of service efficacy, efficiency, and quality (Shell, 2021). The degree to which an organization accomplishes its aims and objectives is another way that Shell (2021) defines organizational performance. According to Kreitner and Kinicki (2022), other academics define organizational performance as the outcomes compared to the strategy.

5 Financial Performance

According to Mazviona et al. (2016), financial performance is the ability of an organization to obtain and utilize its financial resources in various ways in a manner that enables it to develop a competitive advantage. They indicated that excellent financial performance indicates the effectiveness and efficiency of management. Ahokas et al. (2017) analysed financial performance, taking into consideration the viability of the firm. Some of them established the financial performance as the ability of the firm to control the financial resources it has, i.e. its inflows of money should surpass its outflows. To them, performance refers to the ability of the firm to make better use of its financial capabilities, surmount its weaknesses, overcome risks, and embrace opportunities.

Onyali and Okerekeoti (2018) consider the performance listed by two subcategories: accounting-based performance (return on equity, return on assets) and market-based performance (stock price, dividend payout, earnings per share). Whether a firm can make money based on available assets in its core business is subjectively determined by several of the parameters that are associated with firm financial success (Peters & Bagshaw, 2024). In terms of accomplishing set goals, targets, and objectives over a stipulated period, Eyenubo (2021) viewed the aspect of financial performance. These definitions steep the efficiency narrative in the discourse of performance, which resembles that of Ahokas et al. (2017). According to Amoll (2021), the performance is often defined in output terms, where it is discussed as the achievement of quantifiable objectives.

Amoll (2021) reveals that firm performance is a complex architecture that has four elements. The performance of human resources, such as employee satisfaction; the performance rates of the financial and market, such as revenue, profits, market position, cash-to-cash cycle time, and earnings per share; and the effectiveness of the organization, such as the time to market, the level of innovation, and the responsiveness and flexibility of production and supply chains (Valentin, 2014). Lastly, we have the customer-oriented performance, which consists of customer satisfaction and product or service performance.

It is common to distinguish the concept of corporate performance from the considerations of financial matters such as profit, return on assets (ROA), and economic value added (EVA) (Onyali & Okerekeoti, 2018). Financial Performance has been examined by various researchers in a variety of measures. Return on Assets (ROA) was used to measure firm financial performance by Terjesen et al. (2020) and Onyali and Okerekeoti (2018). Financial performance, according to the definition, is a measure of the financial outcome of business operations performed in an organization and its capability to utilize both human and capital assets available to it with ease to advance on performance factors like profitability, liquidity, and efficiency (Ebere et al., 2021).

Loh and Nguyen (2019) have used market capitalization, debt-to-equity ratio, price-to-earnings ratio, return on equity (ROE), and Tobin Q, a mixed metric, to determine the level of financial performance. Whereas Lee-Kuen et al. (2017) applied Tobin Q and considered it to be a measure of business financial performance, Gambo et al. (2021) and Vishwakarma (2017) resorted to Return on Assets (ROA). Although belonging to different spheres, Elmagrhi et al. (2022) selected the dividend payout as an indicator of business financial performance, Abdussalam and Okike (2024) used the Return on Equity (ROE). Both ROA and ROE were used as financial performance indicators by Rao and Desta (2016), whereas the analysis of Muchemwa et al. (2016) represents three basic financial measures of company performance: ROA, ROE, and Tobin Q.

The notion of the financial performance can be estimated in many ways, such as a system of indices of profitability, efficiency, liquidity, shareholder value, and market value of a firm (Kramaric et al., 2020). This kind of measure is explained further in this study. Gambo et al. (2021) cite that the advantage of employing financial performance metrics in studies on diversity and governance is that by doing this, its users can assess the effectiveness of the corporate governance system in place in a company in ensuring and promoting management effectiveness. Financial measures of performance indicate the effectiveness with which resources of a business are utilized by management (Mazviona et al., 2016).

The performance of the company has been measured using return on equity and assets (Obeidat et al., 2017). To measure the performance of the company, Yasser et al. (2021) applied the Profit Margin (PM) and Return on Equity (ROE). Yasser et al. (2017) focused on the business performance computed by Economic Value Added (EVA), market-based Tobin Q, and accounting-based Return on Assets (ROA). Whereas ROE and the Return on Investment (net result plus interest)/(equity plus total debt) were some of the measures used in gauging the financial reporting position, Homan (2017) gauged the position of the company by using the market-based measures, which included the market value of equity divided by book value of equity and Tobin Q (market value of equity plus book value of debt)/(total assets minus in book value).

Bhagat and Black (2022) used Tobin Q, the ratio of the operating income to assets, the ratio of sales and assets, operating margin of the operating income to sales, sales per employee, and growth of assets, sales, operating income, employees, and cash flows as a measure of the dependent variables of company performance. Besides, dividend yield and profitability (return on equity and return on assets) used in the analysis by Mule et al. (2023) indicated the performance of the companies financially. Return on Assets (ROA) is the net income divided by a proportion of total assets. The way it can be broken down is: Where the amount of net income that a company will pay back in proportion to shareholders, post net income, is the Return on Equity (ROE), the EBIT/the average amount of total assets at book value will determine the ROA. The success of a corporation is determined by the amount of money it invests in the investment of shareholders. This is the measure of the corporation to gain returns on its capital through the returns on equity.

Valentin (2014) notes that the historical understanding of financial performance has been purely calculated on its ability to make profits. This has changed with time because the definition of performance has changed depending on the opinion of the user of the financial data. As long as a business can satisfy the interests of all involved parties, including the management interested in their well-being and profit since their contributions are appropriately rewarded and owners who can maximize their wealth by increasing the market value of their companies (which can also be pegged on the profit), then the business can be noted as a global performer.

Kastrati et al. (2020) include in their definition of global performance (ability by the company to share dividends with capital investors considering the risk they undertake; the commercial partners desire the stability and the solvency of the company; the credit institutions request an assurance that the company can reimburse the loans in time (solvency); the employees are interested in having a stable occupation and having significant material benefits of the company; and the state requires a company that is efficient and accountable in paying taxes and creating new workplaces (Valentin, 2014).

Kagzi and Guha (2020) indicate that financial performance analysis can be of great assistance to stakeholders in assessing the performance of the organization in running it. It assists in making complicated numerical explanations and critical analysis easy to interpret. Even though only a portion of the indicators are indeed useful in the financial diagnosis of a company, many researchers use many more indicators instead of what is necessary (Gadoiu, 2023). Moreover, such indices do not represent the absolute standards of assessment, but such parameters could only provide valuable information in combination with other indicators that demonstrate the changes in the operations and financing within the period and in comparison to how the other business in the same market operates (Adeyemi et al., 2022; Gadoiu, 2023).

Business management uses financial indicators to monitor, record, and improve performance. It has been proven that to obtain a global picture of an economic actor at a specific period, it is essential to use a balanced multidimensional system of assessment that uses financial ratios as well as non-financial measures (Al-Karim & Alams, 2023). Based on the above-presented perspectives, the study will consider financial performance as the financial performance of an organization following pre-determined business processes based on the ability of an organization to utilize the available capital and human resources to influence indices such as profitability level, use of liquidity, efficiency, and value of the shareholders in an insurance firm.

6 Methodology

The extraction of data was based on the Preferred Reporting Items of Systematic Reviews and Meta-Analyses (PRISMA), which guarantees the systematic, transparent, and reproducible evidence reporting. PRISMA offers a rigorous procedure without putting at risk the chance of bias in non-systematic reviews (Haddaway et al., 2022) and is rapidly gaining adoption in the management studies as it has the potential to synthesize many-context evidence (Jimoh, 2025). The PRISMA procedure was adopted over four iterative phases: identification, screening, eligibility, and inclusion. Systematic searches were conducted on EBSCOhost, Elsevier, ProQuest, Wiley Online Library, SpringerLink, and Google Scholar databases. Duplicates and unqualifying records were eliminated, leaving 124 full-text studies to be assessed of an original 2645 records. Though they screened many records sourced from citations (2645), no additional studies were eligible through this route. Accordingly, 31 new papers were included. Having evaluated the quality of methods, relevance, and completeness of the data, an additional 14 more papers were included. In total, 138 studies were brought to the final synthesis. This result reflects a comprehensive and methodologically sound review process, clear inclusion/exclusion criteria, complete transparency, and a rigorous attempt to identify all of the available literature.

In synthesizing useful lessons from the diverse sources, the research used a hybrid synthesis strategy that combined thematic synthesis and structured narrative integration. This synthesis was selected to facilitate both qualitative and quantitative data to attain interpretive richness and analytical coherence in diverse studies. Synthesis was carried out in three steps in a systematic manner. First, open coding was used to find out recurring ideas and empirical patterns and created early descriptive categories such as cyber risk vulnerability, information governance, reputational erosion, performance disruption, and adaptive resilience. Second, the connection of these categories in terms of their concepts was done through the use of axial coding that created a relationship situation among technological exposure, governance practices, and performance feedback. Third, a narrow sample of the patterns resulted in four higher-level analytical themes:

Exposure to technology and systemic vulnerability: Studies have generally emphasized that the digitization to increase efficiency creates structural vulnerabilities in data integrity and system reliability. The weaknesses increase vulnerability to cyber risks and operational impediments.

Governance, control, and compliance mechanisms: The results showed that the best cyber governance strategies, awareness of the employees, and compliance processes are essential in avoiding risks, but have been applied inconsistently among institutions and geographic locations.

Reputational and financial impacts of cyber breaches: In all instances, it was found that cyber breaches, loss of stakeholder confidence, and financial poor performance are closely related, especially to high-profile sectors like telecom and banking.

Strategic resilience and organizational learning: Recent advances in the subject matter included the emphasis on adaptive capacity, the investment in the cybersecurity infrastructure, and post-incident learning as the cornerstones of the long-term sustainability and institutional legitimacy.

This multi-layered synthesis made conceptual integration, as opposed to aggregation, easier. The application of synthesis of quantitative findings to impart direction and magnitude of effects, whereas the application of induction of qualitative insights to the richness of context was by coding. The methodological rigour, representativeness of the sample, and theoretical contribution of the evidence were evaluated. It is based on this synthesis that cybersecurity is not a technical issue but an institutional capacity that cuts the relationship between technology adoption and organizational performance. The literature comes together in the realization that in the absence of active governance frames and frequent engagement of employees, technological development heightens systemic vulnerability. Also, reputational loss associated with cyberattacks has a detrimental impact on stakeholder confidence and resilience in the long run.

7 Discussion of Findings

The systematic review of 138 studies according to the PRISMA protocol identified four metathemes explaining a way of how cybersecurity is associated with organizational performance: technological vulnerability, leadership and governance response, reputational impact, and strategic resilience. The metathemes offer a combined structure of the explanation of how cybersecurity relationships operate to influence organizational performance in various spheres. Technological vulnerability was a dominant theme in most of the studies under review. Evidence was always there to prove that faster technological advancement, though enhancing operational efficiency, has also made organizations exposed to new cyber threats. King (2016) and Dunmade et al. (2023) confirm that increased digitalization without related security infrastructure is a factor behind system exposure and operational downtime. Organizations that lacked any security frameworks had repeated incidents that destabilized financial performance and stability. Quite the contrary, organizations that implemented advanced encryption, access control, and data recovery applications realized fewer instances and shorter recovery times. This shows that technological uptake should be coupled with proactive investment in cybersecurity architecture and continuous employee awareness training for the benefit of sustainable performance.

The second theme, leadership and governance response, focused on the importance of organizational structures and leadership of an ethical nature in the response to the challenge of cybersecurity. The studies that were reviewed indicate that organizational resilience is considerably influenced by leadership commitment to obeying and implementing cybersecurity policies. When leaders show accountability and have a culture of security awareness, online safety will be observed by the workers. James and Thomas (2022) reasoned that the governance structures that incorporated the risk management systems and the strategic planning generated more robust cyber defence mechanisms. This observation supports the argument that effective leadership and good governance are conditions for ensuring the integrity of data and trust by stakeholders. Reputation effect was another emergent issue identified within the review. The effects of cyber breaches are very detrimental to the image of the organization, as well as its competitiveness in the long term. Kadiri et al. (2017) felt that small data breaches could undermine consumer confidence and have a negative impact on investor confidence. Synthesis also indicates that loss of reputation usually has more enduring effects compared to concrete financial loss. Organizations whose communication strategies to the public were transparent and carried out sound crisis management recovered faster from losses in reputation. This supports the argument that reputation is a form of intangible property that has a strong impact on organizational performance, and the maintenance of reputation necessitates a strategic approach to managing the relational and digital risk.

The last theme, strategic resilience, was about how organizations reacted to and learnt about cybersecurity threats. As pointed out by Christou (2016) and Boyes et al. (2016), the key factors to determine the resilience of the technological structure are not only the level of technological sophistication but also the ability to anticipate threats, alter processes, and learn from disturbances. Organizations that had invested in training of employees, planning scenarios, and post-incident analysis had high recovery capabilities. The findings also concur with the dynamic capability theory, which postulates that organizations that constantly create internal capabilities and learning organizational systems have the highest chances of surviving external shocks. Strategic resilience thus is an intermediary between technological adaptation and long-term organizational performance. Firstly, the evidence confirms the fact that the organizational performance and cybersecurity are interrelated via various conduits. The review shows that the financial, operational, and reputational performance is reliant on the level at which organizations incorporate technology, leadership, and resilience in their strategic plan. Funny enough, most of the studies involved technical interventions, yet there were fewer studies that had been done concerning the human and cultural aspects of cybersecurity. This weakness highlights a research gap that should be filled in the future to investigate behaviour drivers of shaping cybersecurity activity.

8 Future Research Directions

Future studies ought to examine how cybersecurity threats have been dynamic and evolving, and what they might bring in the long run in terms of organizational performance. The cross-industry/ geographical studies can reveal industry-specific vulnerabilities and resiliency strategies. Moreover, the analysis of the role of new technologies, including artificial intelligence, machine learning, and blockchain, in reducing cybersecurity risks may offer new solutions to the increase of organizational performance. They also suggest that longitudinal studies should be conducted to determine the long-term effects of cybersecurity breaches on financial stability, performance, and reputation. Additional research needs to be conducted on the human factor, such as employee training, insider threats, and organizational culture, as key elements of cybersecurity resilience, in the future. Evaluating the performance of regulatory frameworks and cybersecurity policies in different countries could provide information on the best practices and policy formulation. Lastly, the application of qualitative research techniques (case studies and interviews) in combination with cybersecurity professionals would help to deepen knowledge and develop more comprehensive, strategic interventions to ensure the organization's performance.

9 Recommendations

Given the expenses that businesses are incurring due to cybersecurity, the term "crime wave" is an understatement. Based on the results, the review suggests that developed organizational methods through services are becoming increasingly popular in cybersecurity management and are also very important to all employees and customers. However, given the contemporary climate in which businesses operate, crime is unavoidable. Even though it is impossible to foresee when a crime will occur, research has shown that the costs of prevention and detection have a big impact on how a firm operates. Therefore, the assessment advises businesses to implement more economical practices and more reliable technology that does not allow illicit activities to occur. Banks should invest more in preventive and detection techniques since they are spending a lot of money developing these goods.

The study found that many stakeholders, not just the bank, are impacted in the wake of a cybersecurity incident. According to the assessment, organizations could sign contracts with other financial institutions that offer insurance coverage to insulate against some of the expenses that businesses incur and avoid operations being reduced, which would result in a loss of revenues. According to one study, these expenses are typical for conducting a company and hence have little impact on operations. The results of these studies, however, showed that indirect costs are significantly more important than the other costs, despite this claim. As a result, the analysis suggests that businesses look for legal action and government assistance when creating regulations that define what constitutes cybercrime and what sanctions should be applied to offenders.

An organizational system must, as a matter of importance, have a backup plan or other safeguards in place to deal with any malicious conduct that aims to compromise consumer information, even though vulnerabilities will always arise. With the rapid advancement of technology and information systems worldwide, organizations should not fall behind in terms of security measures and should be vigilant in identifying any authentication and authorization flaws that could compromise confidentiality, availability, and integrity. Online banking service providers must be more attentive to security regulations and provide many layers of protection for online transactions against security risks.

10 Conclusion

For every business to succeed, controlling cybersecurity for organizational performance is essential. According to the study, reputation, organizational effectiveness, and financial performance were all impacted by hacktivism, cybercrime, and cyberespionage. Innovative services and products, as well as the use of new technologies, have affected bank operations in both positive and negative ways. Positively, they have contributed to better customer service delivery, which has increased the banks' bottom line. However, it is also clear and important that these new trends have had a substantial detrimental impact on operations, such as the prevalent and growing vice of cybersecurity and white-collar crime. The analysis of the review shows that the prevention and detection expenses, such as insurance premium costs and IT compliance costs, affect the creation of organizational goods, but not significantly. The other costs contributed significantly to the development of the services due to the presence of the following costs: (a) direct costs, namely, cost of company continuity, direct financial loss, payments of compensation, legal charges and (b) indirect costs, such as harm to the reputation and loss of consumer trust.

References

Abdussalam, F. Y., & Okike, B. M. (2024). A board gender diversity and performance: An evaluation of the Nigerian banking industry. *Journal of Applied Research, 4*(6), 1721–1727.

Adediwura, A., & Tayo, O. (2022). Human capability evaluation approach for cybersecurity in critical industrial infrastructure. *Advances in Intelligent Systems and Computing Advances in Human Factors in Cybersecurity, 501*(10), 169–182.

Adeyemi, S. B., Okpala, O., & Dabor, L. (2022). Factors affecting audit quality in Nigeria. *International Journal of Business and Social Science, 3*(20), 198–210.

Agarwal, J., Osiyevskyy, O., & Feldman, P. M. (2014). Corporate reputation measurement: Alternative factor structures, nomological validity and organisational outcomes. *Journal of Business Ethics, 130*(8), 485–506.

Ahokas, J. (2019). The effect of market orientation as a mediating variable in the relationship between entrepreneurial orientation and SMEs' performance. *Nankai Business Review International, 7*(1), 39–59.

Ahokas, J., Kiiski, T., Malmsten, J., & Ojala, L. (2017). Cybersecurity in ports: A conceptual approach. In W. Kersten, T. Blecker, & C. M. Ringle (Eds.), *Digitalisation in supply chain management and logistics*. Proceedings of the Hamburg International Conference of Logistics (HICL), 23 (pp. 343–359). epubli GmbH.

Ajila, H. (2022). An effective computational technique for the taxonomic position of security vulnerability in software development. *Journal of Computational Science, 25*(5), 388–396.

Ajzen, J., & Fishbein, F. (1980). *Cryptovirology: Extortion-based security threats and countermeasures*. In Proceedings of the IEEE Symposium on Security and Privacy (pp. 129–140).

Ali, A. Y., Pocock, K., & Hu, Q. (2024). The effect of board of directors' IT awareness on CIO compensation and firm performance. *Journal on Decision Sciences, 45*(3), 401–436.

Al-Karim, A., & Alams, R. (2023). Identifying the barriers to Iran's saffron export by using Porter's diamond model. *International Journal of Marketing Studies, 4*(5), 129–138.

Amboko, S., & Wagoki, U. (2022). Facebook loses around $13 billion in value after a data breach affects 50 million of its users. *Fortune*. Retrieved 12th December 2019 from fortune.com/2019/09/28/facebook-stock-falls-after-security-breach

Amoll, D. O. (2021). *Effects of board of directors' composition on the financial performance of companies listed on Nairobi Stock Exchange* (Unpublished Master's thesis). University of Nairobi, Kenya.

Anderson, R. (2004). *Proceedings of a 2004 workshop: Understanding the insider threats*. RAND Corporation.

Anderson, R. (2016). *Security engineering: A guide to building dependable distributed systems*. Wiley.

Barnett, M. L., Jermier, J. M., & Lafferty, B. A. (2016). Corporate reputation: The definitional landscape. *Corporate Reputation Review, 9*(3), 26–38.

Bartholomew, A. (2020). Choosing workers' qualifications: No experience necessary? *International Economic Review, 34*(3), 479–502.
Bawane, W., & Stelke, L. (2019). *Managing diversity in multinational organisation*. Linnaeus University.
Bhagat, C. B., & Black, K. D. (2022). Us versus them: The roles of organizational identification and disidentification in social marketing initiatives. *Journal of Public Policy & Marketing, 21*(1), 26–36.
Bhorn, G. M. (2021). Concepts and theory of organization-public relationships. In J. A. Ledingham & S. D. Bruning (Eds.), *Public relations as relationship management: A relational approach to the study and practice of public relations* (pp. 3–22). Lawrence Erlbaum Associates.
Biener, C., Eling, M., & Wirfs, J. H. (2021). Insurability of cyber risks: An empirical analysis. *The Geneva Papers on Risk and Insurance—Issues and Practice, 40*(1), 131–158.
Bikker, C. (2023). Crisis management: Advantages of a relational perspective. In J. A. Ledingham & S. D. Bruning (Eds.), *Public relations as relationship management: A relational approach to the study and practice of public relations* (pp. 73–94). Lawrence Erlbaum Associates.
Borum, R., Felker, J., Kern, S., & Dennesen, T. F. (2022). Strategic cyber intelligence. *Information & Computer Science, 23*(3), 317–332.
Boyes, H., Isbell, R., & Luck, A. (2016). *Code of practice: Cyber security for ports and ports systems*. The Institution of Engineering and Technology (IET).
Brenner, J. (2022). The new industrial espionage. *Journal of Management, 8*(2), 28–29. Retrieved 16th December 2024 from http://www.theamerican-interest.com
Broadhurst, R. (2020). Developments in the global law enforcement of cyber-crime Policing. *An International Journal of Police Strategies and Management, 29*(2), 408–433.
Canavan, J. (2018). *Fundamentals of network security*. Artech House, Inc.
Carrapico, H., & Barrinha, A. (2022a). The EU as a coherent cyber (security) actor. *Journal of Common Market Studies, 55*(6), 1–9.
Carrapico, I., & Barrinha, D. (2022b). *A large-scale empirical study of security patches*. University of California, Berkeley and International Computer Science Institute. Retrieved 12th December 2024 from http://people.eecs.berkeley.edu
Chang, S. C., & Ho, N. P. (2019). Aligning manufacturing strategy with business strategy: An empirical study in high-tech industry. *International Journal of Technology Management, 24*(8), 70–87.
Choi, S., & Chan, H. G. (2022). Managing diversity in US federal agencies: Effects of diversity and diversity management on employee perceptions of organizational performance. *Public Administration Review, 70*(1), 109–121.
Christou, G. (2016). *Cybersecurity in the European Union: Resilience and adaptability in governance policy*. Palgrave and Macmillan.
Chun, R. (2021). Corporate reputation: Meaning and measurement. *International Journal of Management Reviews, 7*(2), 91–109.
Cole, E. (2006). *Insider threats: Protecting the enterprise from sabotage, spying and theft*. Syngress Publishing.
Debreu, C. (2023). The effects of higher education policy on the location decision of individuals: Evidence from Florida's Bright Futures Scholarship Program. *Regional Science and Urban Economics, 39*(2), 553–562.
Drover, W., Wood, M. S., & Fassin, Y. (2021). Take the money or run? Investors' ethical reputation and entrepreneurs' willingness to partner. *Journal of Business Venturing, 29*(6), 723–740.
Dunmade, E. O., Ajayi, M. A., & Jimoh, A. L. (2023). Globalization and its impact on human resource management practices in Pz Cussons Nigeria Plc., Lagos. *The Journal of Business Studies, 7*(1). Retrieved 16th December 2024 from https://www.fcm.esn.ac.lk/jbs/vol_07-1.html
Ebere, C. C., Ibanichuka, E. A. L., & Ogbonna, G. N. (2021). Corporate governance system and financial performance of quoted insurance companies in Nigeria. *International Journal of Business & Law Research, 4*(4), 34–41.

Eling, M., & Luhnen, S. (2016). When does market share matter? New empirical generalizations from a meta-analysis of the market share-performance relationship. *Journal of Marketing, 82*(4), 1–24.
Elmagrhi, M. H., Ntim, C. G., Crossley, R., Malaglia, J. K., Fosu, S., & Vu, T. V. (2022). Dividend pay-out policy in UK listed SMEs: The effect of corporate board characteristics. *International Journal of Accounting and Information Management, 5*(2), 1–34.
Elnaga, A., & Imran, A. (2018). *The effect of training on employee performance.* The International Institute for Science, Technology and Education.
Ernst & Young. (2018). *Viewpoints issue 19; 2012.* Retrieved 16th December 2024 from http://www.ey.com/publication/vwLUAssets/ACLS_viewpoints19.pdf
Eyenubo, S. (2021). Chief executive officer duality and financial performance of firms in Nigeria. *International Journal of Business & Social Research, 2*(6), 125–134.
Farell, M. (2021). On the security of today's online electronic banking systems. *Computers & Security, 213*(3), 253–265.
Fitzgerald, M., Kruschwitz, N., Bonnet, D., & Welch, M. (2020). Embracing digital technology: A new strategic imperative. *MIT Sloan Management Review, 55*(2), 1–12.
Fombrun, C. J., & Van Riel, C. B. M. (2021). The reputational landscape. *Corporate Reputation Review, 1*(1/2), 5–13.
Gadoiu, M. (2023). Advantages and limitations of the financial ratios used in the financial diagnosis of the enterprise. *Scientific Bulletin Economic Sciences, 13*(2), 87–94.
Gambo, E. J., Bello, B. A., & Rimamshung, S. A. (2021). Effect of board size, board composition and board meetings on financial performance of listed consumer goods in Nigeria. *International Business Research Journal, 11*(6), 1–10.
Gennaro, A., Vineet, M., & Scott, W. (2022). Understanding nonmalicious security violations in the workplace: A composite behavior model. *Journal of Management Information Systems, 28*(2), 203–236.
Gerald, O. (2020). Cyber-espionage: A growing threat to the American Economy. *Journal of Communication & Poly, 24*(2), 241–242.
Gercke, M. (2022). The slow wake of a global approach against cybercrime. *Computer Law Review International, 2*(2), 141–156.
Global Risks Report. (2017, January 11). *World Economic Forum.* Global Risks Report.
Gray, E. R., & Balmer, J. M. (2022). Managing corporate image and corporate reputation. *Long Range Planning, 31*(5), 695–702.
Griffel, S., & Lovell, D. (2017). Information systems security policy compliance: An empirical study of the effects of socialisation, influence, and cognition. *Information & Management, 51*(1), 69–79.
Gross, M., Canetti, D., & Vashdi, D. (2020). Cyberterrorism: Its effects on psychological well-being, public confidence and political attitudes. *Journal of Cybersecurity, 3*(1), 49–58.
Gurses, M., & Kilic, M. (2019). The effects of board gender diversity on firm performance: Evidence from Turkey. *International Journal of Business and Management, 10*(9), 182–192.
Haddaway, N. R., Page, M. J., Pritchard, C. C., & McGuinness, L. A. (2022). PRISMA2020: An R package and Shiny app for producing PRISMA 2020-compliant flow diagrams, with interactivity for optimised digital transparency and Open Synthesis. *Campbell Systematic Reviews, 18*, e1230. https://doi.org/10.1002/cl2.1230
Heather, K. (2013, August 1, 10:35 AM). *NSA chief: Snooping is crucial to fighting terrorism.* CNN. Retrieved 23rd December 2024 from http://www.cnn.com
Hedayati, A. (2022). An analysis of identity theft: Motives, related frauds, techniques and prevention. *Journal of Law and Conflict Resolution, 4*(1), 1–12.
Herbig, P., & Milewicz, J. (2021). The relationship of reputation and credibility to brand success. *Journal of Consumer Marketing, 12*(1), 5–10.
Homan, R. (2017). *Dutch board diversity and firm performance* (Unpublished Master's Thesis). University of Groningen, Netherlands.

Howlett, W. (2019). *The rise of China's hacking culture: defining Chinese hackers*. Retrieved 12th December 2024 from http://scholarworks.lib.csusb.edu/etd

Ibikunle, T., & Eweniyi, R. (2020). Customer loyalty and electronic banking: A conceptual framework. *Journal of Global Information Management, 9*(3), 11–23.

International Telecommunication Union (ITU). (2018). *Global cybersecurity index 2018: Measuring commitment to cybersecurity*. Geneva, Switzerland: ITU Publications.

Investopedia. (2021). *Trust in cyberspace*. Committee on Information Systems Trustworthiness, National Research Council (USA). Retrieved 23rd August 2024 from http://investopedia.edu/books/0309065585/html/index.html

ISO/IEC 27000. (2016). *Risk management Principles and guidelines on implementation*. International Organization for Standardization. Retrieved 29 November 2024 from www.iso.org

James, D. U., & Thomas, M. M. (2022). *Cybersecurity for manufacturers: Securing the digitized and connected and connected factory*. MForesight Publication.

Jassal, K., & Sehgal, J. (2023). Information privacy in cyberspace transactions. *Stanford Law Review, 50*(4), 1193–1998.

Jimoh, A. L. (2025). Talent management practice in the contemporaneity organization. A systematic literature review. *Economic Insights—Trends and Challenges, 14*(1), 59–72. https://doi.org/10.51865/EITC.2025.01.04

Jimoh, A. L., Dunmade. E. O., & Salman, A. (2025). Green talent management and job performance: the significance of open innovation. *Journal of Intellectual Capital*, 1–18. https://doi.org/10.1108/JIC-03-2025-0069

Jung, P. W. (2019). A critical analysis on the concept of "cyber security." *Yonsei Journal of Medical and Science Technology Law, 2*(2), 1–25.

Kadiri, I. B., Isiaka S. B., & Jimoh, A. L. (2017). Relationship between talent management and labour turnover in Nigeria banking sector. *Journal of Humanities, 2*(2). Retrieved 16th December 2024 from https://www.ijhumas.com/ojs/index.php/kiuhums/issue/view/8

Kagzi, M., & Guha, M. (2020). Board demographic diversity: A review of literature. *Journal of Strategy and Management, 11*(1), 33–51.

Kamel, S. (2020). The use of information technology to transform the banking sector in developing nations. *Information Technology for Development, 11*(4), 305–312.

Kapto, A. S. (2023). Cyberwarfare: Genesis and doctrinal outlines. *Herald of the Russian Academy of Science, 83*(4), 357–364.

Kastrati, A., Adnett, N., & Toci, V. (2020). The relationship between the output gap and excess liquidity: Evidence from Czech Republic, Estonia and Kosovo. *Journal of Economics and Management, 31*(1), 95–109.

Kendrick, R. (2017). *Cyber risks for business professionals: A management guide*. IT Governance Publishing.

King, T. (2016). Information systems security policy compliance: An empirical study of the effects of socialisation, influence, and cognition. *Information & Management, 51*(1), 69–79.

Kitheka, P. M. (2019). Information security management systems in public universities in selected institutions. *Strategic Management Journal, 37*(3), 466–476.

Kitheka, P. M. (2023). *Information security management systems in public universities in Kenya: A gap analysis between common practices and industry best practices*. University of Nairobi.

Kong, E., & Farrell, M. (2020). The role of image and reputation as intangible resources in nonprofit organisations: A relationship management perspective. *Journal of International Studies, 10*(4), 207–218.

Kramaric, T. P., Aleksic, A., & Pejie-Bach, M. (2020). Measuring the impact of board characteristics on the performance of Croatian insurance companies. *International Journal of Engineering and Business Management, 10*(1), 1–13.

Kreitner, R., & Kinicki, A. (2022). *Organizational behaviour*. McGraw-Hill.

Lange, D., Lee, P. M., & Dai, Y. (2021). Organizational reputation: A review. *Journal of Management, 37*(1), 153–184.

Lee-Kuen, I. Y., Sok-Gee, C., & Zairudin, R. (2017). Gender diversity and firm financial performance in Malaysia. *Asian Academy of Management Journal of Accounting and Finance, 13*(1), 41–62.

Lindskog, S. (2016). *Observations on operating system security vulnerabilities*. Lincentiate of Engineering.

Loh, L., & Nguyen, M. H. (2019). Board diversity and business performance in Singapore listed companies: The role of corporate governance. *International Journal Research Publication, 7*(10), 95–104.

Loonam, M., & O'Loughlin, D. (2021). An observation analysis of e-serve quality in online banking. *Journal of Financial Services Marketing, 13*(2), 164–178.

Luppicini, R. (2024). Illuminating the dark side of the internet with actor network theory: An integrative review of current cybercrime research. *Global Media Journal, Canadian Edition, 7*(1), 35–49.

Magutu, P. O., Ondimu, G. M., & Ipu, C. J. (2021). Effects of cybercrime on state security: Types, impact and mitigations with the fibre optic deployment in Kenya. *Journal of Information Assurance & Cyber Security, 2*(1), 67–74.

Mathis, R. L., & Jackson, J. H. (2017). *Human resource management* (12th ed.). Thompson South Western.

Maurushat, A. (2018). *Disclosure of security vulnerabilities: Legal and ethical issues*. Springer Publication.

Mazviona, B. W., Dube, M., & Sakahuhwa, T. (2016). An analysis of factors affecting the performance of insurance companies in Zimbabwe. *Journal of Finance and Investment Analysis, 1*(1), 1–10.

Moses-Oke, R. O. (2019). Cyber capacity without cyber security: A case study of Nigeria's national policy for information technology (NPFIT). *The Journal of Philosophy, Science & Law, 12*(3). Retrieved 16th December 2024 from www.Miami.Edu/Ethics/Jpsl

Muccin, E. (2022). Combatting maritime cyber security threats. *Maritime Reporter and Engineering News, 77*(6), 43–52.

Muchemwa, R., Callaghon, C. W., & Padia, N. (2016). Board composition, board size and financial performance of Johannesburg stock exchange companies. *South African Journal of Economic and Management Science, 19*(4), 497–513.

Mule, R. K., Mukras, M. S., & Nzioka, O. M. (2023). Corporate size, profitability and market value: An econometric panel analysis of listed firms in Kenya. *European Scientific Journal, 11*(13), 376–398.

NESA. (2018). *The Finnish national emergency supply agency* (NESA7). Retrieved 16th December 2024 from https://www.nesa.fi

Nwankwo, R. (2020). Cybercrime in Nigeria. *Business Intelligence Journal*. Retrieved 16th December 2024 from http://www.saycocorporativo.com/sayco.pdf

Obeidat, B., Khasawneh, A., & Altal, A. (2017). The impact of knowledge management on innovation: An empirical study on Jordanian consultancy firms. *Management Research Review, 39*(3), 1214–1238.

Odhiambo, R. (2018). Beyond strategic information systems: Towards an IS capability. *Journal of Strategic Information Systems, 13*(2), 167–194.

Odunfa, A. (2021). *Nigeria: Report on cyber threat calls for quick passage of 2012 bill*. Retrieved 23rd December 2024 from http://www.allafrica.com/stories/201405080279.Html

Ogut, H., Raghunathan, S., & Menon, N. (2022). Cyber security risk management: Public policy implications of correlated risk, imperfect ability to prove loss, and observability of self-protection. *Risk Analysis, 31*(3), 497–512.

Ojeka, S. A., Ben-Caleb, B., & Ekpe, E. I. (2017). Cyber security in the Nigeria banking sector: An appraisal of audit committee effectiveness. *International Review of Management and Marketing, 7*(2), 340–346.

Ojo, V. (2021). An appraisal of the practice of corporate diversification in selected Nigerian banks. *Management Science Letters, 1*(1), 1295–1306.

Onyali, C. I., & Okerekeoti, C. U. (2018). Board heterogeneity and corporate performance of firms in Nigeria. *International Journal of Academic Research in Accounting, Finance and Management Sciences, 8*(3), 103–117.

Oparanma, A. O. (2021). The organizational culture and corporate performance in Nigeria. *International Journal of African Studies, 3*(3), 34–40.

Pappalardo, J. (2023). NSA data mining: How it work. *Popular Mechanics*. Retrieved 12th December 2024 from http://www.popularmechanics.com/military/a9465/nsa-data-mining-how-it-works-15910146

Peters, G. T., & Bagshaw, K. B. (2024). Corporate governance mechanism and financial performance of listed firms in Nigeria: A content analysis. *GlobalJournal of Contemporary Research in Accounting, Auditing and Business Ethics, 1*(2), 103–128.

Polemi, N. (2018). *Port cybersecurity: Securing critical information infrastructures and supply chains*. Elsevier.

Raggad, B. G. (2010). *Information security management: Concepts and practice*. CRC Press.

Raggad, B. G. (2018). Do information security professionals and business managers view information security issues differently? *Information Systems Security, 16*(2), 100–108.

Rantapelkonen, J., & Kantola, H. (2018). Insights into cyberspace, cyber security, and cyberwar in the Nordic Countries. In J. Rantapelkonen & M. Salminen (Eds.), *The fog of cyber defense* (pp. 24–36). National Defense University.

Rao, K. S., & Desta, K. K. (2016). Corporate governance and financial performance: A study with to commercial banks in Ethiopia. *International Journal of Applied Research, 2*(8), 551–557.

Rathaus, N. (2019). *Vulnerability assessment white paper: Automating vulnerability assessment*. Retrieved 11th August 2024 from www.securiteam.com

Reid, R., & Van Niekerk, J. (2014). *From information security to cyber security cultures*. In Proceedings of the 13th ISSA annual meeting. Information security for South Africa (pp. 1–7). IEEE.

Richardson, O. C. (2019). Racial diversity, business strategy, and firm performance: A resource-based view. *Academy of Management Journal, 43*(2), 164–177.

Rosewarne, S. (2019). Migrant domestic work: From precarious to precarisation. *Journal fuer Entwicklungspolitik/Austrian Journal of Development Studies, 30*(4), 133–154.

Safa, N. S., Von Solms, R., & Furnell, S. (2021). Information security policy compliance model in organizations. *Computers & Security, 56*(6), 70–82.

Schultz, E. (2022). A framework for understanding and predicting insider attacks. *Computer & Security, 21*(6), 525–531.

Shell, K. (2021). The evolving landscape of maritime cybersecurity. *Review of Business, Saint John's University, 25*(3), 30–36.

Siddique, I., & Rehman, S. (2017). Impact of electronic crime in Indian banking sector—An overview. *International Journal of Business & Information Technology, 1*(2), 60–71.

Song, J., & Windram, B. (2018). *The effectiveness of audit committee: experience from UK* (Working Paper). 12th Asian-Pacific Conference on International Accounting Issues. Beijing, China, pp. 21–24.

Sterlin, R. (2017). The evolving landscape of maritime cybersecurity. *Review of Business, Saint John's University, 25*(3), 30–36.

Stoner, J. A. F., Freeman, R. E., & Gilbert, Jr., D. R. (2016). *Management* (6th ed). Prentice Hall of India Private Limited.

Suleiman, A., & Olayinka, T. (2021). Cybersecurity challenges and information protection in digital environments. *International Journal of Cybersecurity and Information Systems, 8*(2), 45–59.

Tanenbaum, S. (2011). *Computer networks* (4th ed.). Published by Pearson Education & Dorling Kindersley.

Teoh, C. S., & Mahmood, A. K. (2017). *National cyber security strategies for digital economy*. 5th International Conference on Research and Innovation in Information Systems (ICRIIS), Langkawi, Malaysia.

Terjesen, S., Conto, E. B., & Francisco, P. M. (2020). Does the presence of independent and female directors impact firm performance? A multi-country study of board diversity. *Journal of Management and Governance, 20*(3), 447–483.

Tonge, A. M., Kasture, S. S., & Chaudhari, S. R. (2019). Cyber security: Challenges for society—Literature review. *IOSR Journal of Computer Engineering, 12*(2), 67–75.

Tucci, A. (2017). Cyber risk management preparing for new operational risks. *Port Technology International (PTI) Journal, 74*(5), 90–92.

Valentin, E. K. (2014). *Business planning and market strategy*. Sage.

van Der Wiele, D. (2016). Cybersecurity: The environment of the banking industry. *Journal of Issues in Information Systems, 16*(3), 181–200.

Venkatraman, N., & Ramanujam, V. (2016). Measurement of business economic performance: An examination of method convergence. *Journal of Management Development, 13*(1), 109–122.

Vishwakarma, R. (2017). Women on board and its impact on performance: Evidence from microfinance sector. *Indian Journal of Corporate Governance, 10*(1), 58–73.

Von Solms, R., & Van Niekerk, J. (2018). From information security to cyber security. *Computers & Security, 38*, 97–102.

Wada, F., Longe, O., & Danquah, P. (2022). Action speaks louder than words—Understanding cyber-criminal behaviour using criminological theories. *Journal of Internet Banking and Commerce, 17*(1), 343–359.

Wasilwa, K. (2019). Establishing a theory of cybercrimes. *International Journal of Cyber Criminology, 1*(2), 7–9.

Weisenburger, K. (2021). *Hacktivists of the world, divide*. Retrieved 18th July 2024 from http://www.SecurityWatch.com

Whitman, M., & Mattord, H. (2019). *Principles of information security* (4th ed.). Cengage Learning.

Whyte, P. (2022). ECDIS-Navigation in 2021. *Port Technology International (PTI} Journal, 77*(2), 104–105.

Wokoma, C. U., & Iheriohanma, E. B. J. (2021). Interaction between globalization and organizational performance in the third world: Nigeria. *Studies in Sociology of Science, 7*(1), 72–80.

Yasser, Q. R., Al Mamum, A., & Rodrigs, M. (2017). Impact of board structure on firm performance: Evidence from an emerging economy. *Journal of Asia Business Studies, 11*(2), 210–228.

Yasser, Q. R., Entebang, H., & Al Mamun, A. (2021). Corporate governance and financial performance in Pakistan—The case of Karachi stock exchange. *Journal of Economics and International Finance, 3*(8), 482–491.

Yassir, A., & Nayak, S. (2022). Cybercrime: A threat to network security. *International Journal of Computer Science and Network Security, 12*(2), 91–97.

Ibrahim Olanrewaju Lawal is an accomplished Business Lecturer currently based at the GBS/University of Suffolk's Birmingham Campus in the United Kingdom (UK). With over a decade of experience in higher education across two countries, he specializes in Business, Human Resource Management, Organisation behaviour, and Management studies. He is recognized as a Fellow of the Higher Education Academy (FHEA), United Kingdom, and a Certified Management & Business Educator (CMBE), United Kingdom, reflecting his commitment to excellence in teaching and professional standards. He holds a PhD in Management from the University of KwaZulu-Natal in South Africa, as well as an MSc and BSc in Human Resources and Industrial Relations from Lagos State University, Nigeria. Before his current role in the UK, he served for more than ten years at Kwara State University (KWASU) in Nigeria, where he progressed to the position of Senior Lecturer and held various administrative leadership roles, including Coordinator for the Professional Degree Programme and Examination Officer.

Evwiekpamare Fidelis Olori is a Lecturer at Global Banking School, UK, with extensive experience in teaching management, business, finance, and entrepreneurship across UK higher education. He holds a PhD in Public Administration, complemented by five master's degrees across various areas within management and the social sciences, as well as a Postgraduate Certificate in Higher Education. His research spans Public Administration, Governance, Public Policy, Management, Business, Entrepreneurship, and Higher Education, with a growing portfolio of peer-reviewed publications in these areas. He has a proven record of delivering high-quality teaching, curriculum development, and student engagement across diverse learning environments. Dr Olori is a Fellow of Advance HE and a Chartered Member of the Chartered Association of Business Schools. He combines academic scholarship with professional expertise in public sector management, policy analysis, business management, and organizational leadership.

Ismaila Yusuf is an Associate Professor in the Department of Business and Entrepreneurship at Kwara State University (KWASU), where he currently serves as the Acting Head of Department. He holds a PhD and M.Sc. in Business Administration from the University of Ilorin and a B.Sc. in Management Studies from Usmanu Danfodiyo University, Sokoto. His academic work is characterized by a strong interdisciplinary focus on Human Resource Management, Strategic Management, and Entrepreneurship, emphasizing sustainable business practices and employee engagement. Professionally, he is recognized as a Fellow of the Chartered Institute of Public Diplomacy and Management (CIPDM) and an Associate Member of the Institute of Strategic Management Nigeria (ISMN). Yusuf has published extensively in both international and national journals on topics including conflict management, SME survival, and corporate performance. His scholarly influence extends to postgraduate supervision, where he has guided numerous PhD and master's students, and he serves as an External Examiner for doctoral dissertations at UNICAF University. Beyond academia, he is deeply committed to community service, frequently serving as a supervisor for national examinations and leading local initiatives focused on economic empowerment and educational support.

Ethical Dilemmas of Generative AI and Large Language Models in Human Resource Management

Brian Mabuyana, David Mhlanga, Mufaro Dzingirai, and Emmanuel Ndhlovu

Abstract The rapid technological advancements witnessed in the early twenty-first century are reforming human resource management at a faster pace. However, there is scant evidence pertaining to the influence of generative Artificial Intelligence (AI) and Large Language Models (LLMs) on human resource management. This chapter seeks to bridge this gap. This chapter aims to examine how generative AI and LLMs influence talent management in a data-oriented context. A desktop approach and a conceptual approach were used as the data collection technique. The results revealed that making use of generative AI and LLMs alters the human resource management within a data-oriented setting. On the other side, the results showed that risks and ethical dilemmas affect the use of generative AI and LLMs within the data-oriented workplace. These results are useful for human resource practitioners, policymakers, and scholars.

B. Mabuyana (✉)
Department of Marketing and Retail Management, University of South Africa, Pretoria, South Africa
e-mail: mabuyb@unisa.ac.za

D. Mhlanga
Monash University, Clayton, VIC, Australia
e-mail: david.mhlanga@monash.edu

M. Dzingirai
Harold Pupkewitz Graduate School of Business, Namibia University of Science and Technology, Windhoek, Namibia

Faculty of Business and Communications, INTI International University, Nilai, Malaysia

M. Dzingirai
e-mail: mdzingirai@nust.na

E. Ndhlovu
School of Public Management, Governance and Public Policy, College of Business and Economics, University of Johannesburg, Johannesburg, South Africa
e-mail: endhlovu@uj.ac.za

M. Dzingirai and S. O. Atiku (eds.), *Strategic Human Resource Analytics and Intelligent Data-Driven Systems*, https://doi.org/10.1007/978-981-92-2623-8_10

Keywords Generative Artificial Intelligence · People Analytics · Human Resource Management · Fourth Industrial Revolution · Talent Management · Large Language Models

1 Introduction

Generative Artificial Intelligence (AI) and Large Language Models (LLMS) are transforming the Human Resource Management (HRM) (Barocas et al., 2023; Bhalla & Suribabu, 2024). This transformation has been realized by the HRM processes that operate automatically and amplify metrics-based data. Generative AI and LLMs are critical and efficient as they improve HRM's strategic planning through predictive modelling, customized training material, tailored educational experiences and accurate feedback (Farinha et al., 2025). Therefore, this amalgamation changes the focus of HRM from being administrative to a purpose-driven human development. The use of generative AI and LLMs has shifted from being abstract to a conventional business and user-friendly tool for HRM. In the same vein, the adoption of generative AI and LLMs has been witnessed in various organizations globally. For example, Porkodi and Cedro (2025) aver that nearly 78% of organizations with HRM are using generative AI. In addition, most of the European organizations are using generative AI for their HRM functions while 67% of the organization globally have integrated LLMs so as to support the generative AI operations (Ruokenen, 2025; Barocas et al. 2023). Furthermore, LLM applications are estimated to yield $750 million before 2030 and the adoption of generative AI has been faster than the internet/personal computers at their development stage (Worek & Aaltonen, 2025).

Empirical evidence from previous studies demonstrates that organizations benefit when they incorporate generative AI and LLMs into HRM (Budhwar et al., 2023; Cheong et al., 2024; April & Daya, 2025). Based on the findings of these previous studies, the benefits of adopting the above-mentioned technologies include efficiency, personalized employee experience, decision-making, and enhanced ethical concerns. In the same vein, generative AI and LLMs, when incorporated into recruitment and hiring, improve accuracy and inclusivity when it comes to job descriptions (Ajayi et al., 2024; Chen, 2023). In addition, these technologies have the capabilities of screening and matching candidates effectively as they go beyond matching the keywords to recognize the context and details of the curriculum vitae (Wang & Pashmforoosh, 2024). Generally, this process may assist in the minimization of unconscious bias. On the other side, generative AI is able to streamline communication through scheduling interviews or sending rejections to candidates automatically. Relatedly, LLMs have the ability to personalize interview questions based on the candidate's background. Thus, this process saves time and improves candidate experience.

Another area of study examined the advantages of using AI and LLMs in employee training and development within HR. These studies found that generative AI is able to transform the HR sector because of its ability to upskill and reskill employees

(Amoah, 2024; Du, 2024). Some scholars hold the notion that LLMs generate personalized learning avenues and training content that is focused on employees' career goals, skills, and learning styles. In addition, LLMs are able to create practical simulations tailored to the training of employees in a safe and controlled environment. On the other hand, LLMs provide ongoing feedback and support due to their ability to analyse data performance (Li, 2024; Raghavan et al., 2020). However, there is scant evidence pertaining to the risks associated with generative AI and LLMs. Hence, this chapter seeks to fill in this gap.

The purpose of this chapter is to identify the advantages as well as the disadvantages associated with the use of generative AI and LLMs in HRM. This conceptual chapter draws from a review of primary and secondary sources. The chapter reviews academic documents, including books, book chapters, and journal articles. It also reviews grey literature, including organizational reports, mainly by reputable research and policy organizations, including the World Bank, World Health Organization (WHO), and United Nations Educational, Scientific and Cultural Organization (UNESCO). It also reviews opinion documents such as newspaper articles and blogs. Although offering low chances of replicability, the conceptual approach adopted in this study was critical in clarifying complex and ill-understood benefits of generative AI and LLMs, particularly in HRM. The chapter focuses on the risks associated with generative AI and LLMs which are critical for understanding how the identified risks can be managed within HRM. This chapter assists human resource managers, academics, and policymakers in understanding the importance of generative AI and LLMs as they are incorporated into HRM.

The chapter is structured as follows: Second section discusses the risks and ethical dilemmas of the generative AI and LLMS. This is followed by an evaluation of the outcomes of integrating the above-mentioned tools for Africa and other developing economies. Thereafter, recommendations are drawn from the review and the discussion. Lastly, conclusions are drawn from the review and discussion.

2 Risks and Ethical Dilemmas

Human Resource Management (HRM) faces several risks and moral challenges when generative AI and (LLMs) are incorporated (Budhwar et al., 2023). Although these technologies can enhance decision-making and streamline processes, they also present significant challenges regarding privacy, human agency, bias, and legal compliance (Bhalla & Suribabu, 2024).

Table 1 outlines the risks and ethical implications linked to the adoption of generative AI and LLMs in HRM. These implications include privacy concerns, overreliance on cutting-edge technologies, data bias, and statutory requirements of HRM.

Table 1 Ethical Dilemmas of Generative AI and LLMs in HRM

Risk / ethical dilemma	Description	Key references
Biases in training data and outputs	Generative AI models can inherit historical biases from training datasets, leading to discriminatory recruitment, promotion, and performance evaluation outcomes. For example, biased hiring tools may disadvantage specific genders or ethnic groups.	Barocas et al. (2023); Dastin (2018); Mehrabi et al. (2021); Raghavan et al. (2020)
Privacy and surveillance concerns	AI-driven HR tools often collect and analyse sensitive employee data, raising privacy issues and potential violations of data protection laws. Overreliance on monitoring may lessen trust which may lead to a culture of strict monitoring in the workplace.	Zuboff (2019); Ball (2021); European Parliament (2016)
Overdependence on automation and reduced human agency	The tendency to favour AI outputs over human judgement can result in biased decisions. Thus, human decision-makers rely on AI outputs without critical review.	Bryson (2018); Shrestha et al. (2019)
Legal and compliance issues in HR contexts	The use of AI in HRM has to abide with inclusive practices, employment laws, and data protection laws. Failure to abide with the above-mentioned laws will result in legal action against the organization.	European Parliament (2016); Floridi et al. (2018)

2.1 Biases in Training Data and Outputs

Large Language Models (LLMs) and other generative AI systems are trained on enormous datasets that inevitably contain sociological and historical biases. When incorporated into AI systems, these biases which may be based on gender, race, ethnicity, socioeconomic status, or other demographic traits can affect HRM outcomes like hiring, advancement, and performance reviews (Du, 2024; Mehrabi et al., 2021). For instance, in hiring, if training data indicates a prior preference for applicants from particular backgrounds, the AI may give preference to similar profiles, thereby maintaining rather than eradicating disparities (Chen, 2023). One well-known instance is the AI resume screening tool that Amazon developed. The model began to penalize applications that contained terms related to women, such as "women's chess club" or graduates from women's colleges, because it was trained on resumes submitted to the company over a ten-year period, the majority of which were from male candidates (Dastin, 2018). This example demonstrates how systemic discrimination can be sustained by the careless use of biased training data. Data proxies can also reveal indirect biases. Correlations between race or socioeconomic status and postal codes, educational institutions, or particular career histories can lead to indirect discrimination even in the absence of explicit demographic characteristics (Raghavan et al., 2020). To solve these problems, multi-layered interventions are required. Among these are the use of algorithmic fairness metrics, regular bias audits, and transparency during the preparation and data selection stages (Barocas et al. 2023). Identifying and minimizing unintentional biases can also be facilitated by involving a variety of stakeholders in the design and evaluation of AI, including HR experts, ethicists, and affected employees. Finally, if deliberate precautions are not taken, the use of generative AI in HR may compromise diversity and inclusion initiatives. It is essential to make sure AI technologies are effective and equitable in order to support fair and legally permissible HR practices in the era of automation.

2.2 Privacy and Surveillance Concerns

To use generative AI and LLMs in HRM, a lot of employee data must be collected, stored, and analysed, including demographic profiles, personal identifiers, communication histories, and behavioural patterns (Zuboff, 2019). While these abilities can help with decision-making, they also raise significant concerns about consent, privacy, and the ethical limits of workplace monitoring. For example, AI-powered solutions can monitor the focus of remote workers using webcams, track keystrokes, record application usage time, and assess employee emails for sentiment. Although this kind of monitoring might boost output, it runs the risk of creating a surveillance-focused workplace that erodes trust between employees and employers (Ball, 2021). In addition, there are legal implications that result from the misuse of the data. Relatedly, there are data protection laws such as the General Data Protection Regulation that impose strict limitations on the processing of personal data, particularly where automated decision-making is involved (European Parliament, 2016). The current regulation that applies to data privacy in the EU points out that workers are entitled to transparency, informed consent, and the right to contest decisions made solely by automated systems. Similar privacy legislation, like the California Consumer Privacy Act (CCPA), provides comparable protections in other jurisdictions (Lancieri, 2022). Beyond compliance, privacy concerns are intimately related to organizational ethics (Stahl & Wright, 2018). Stress, decreased job satisfaction, and abuse of workers are a result of intense watchfulness (Ball, 2021). In addition, organizations may be susceptible to risk as a result of non-observance of the use of sensitive HRM information. Further, Companies can face liabilities and damage to their company reputation because of data breaches or because of the use of private HR records (Spinello, 2021). Companies should adopt the privacy-by-design principles, limit data collection to only what's needed, and keep communication with the staff members about the usage of staff data. The privacy-by-design approach cuts those risks (Ebert et al. 2021). Crucially, taking employee input into account when creating policies can aid in striking a balance between business objectives and the basic right to privacy at work.

2.3 Overdependence on Automation and Reduced Human Agency

Despite the fact that generative AI could enhance productivity and ensure objective-based accuracy of decision-making, dependence on these systems for HR decisions may lead to diminution of human agency and judgement (Porkodi & Cedro, 2025). According to Shrestha et al. (2019), "automation bias" arises when human decision-makers blindly heed AI advice over and above any conflicting evidence-based information. Taking the rank lists or performance scores generated by these AI solutions without a thoroughly conducted human judgement might demonstrate bias in HR settings, as this overlooks contextual factors influencing performance, for example, the learning potential of the individual, team fit, and unique situation (April & Daya, 2025). HR practices may start to become more inhumane as there is a reduction in human agency. Bryson (2018) states that anywhere it would be impossible to realize discretion, empathy, and ethical judgement that is required in complex personnel decisions such as promoting employees, terminating employees, or making disciplinary action decisions. Where a human manager would potentially find mitigating circumstances such as illness, caring responsibilities, and organizational barriers. Problems with the system that controls how individuals complete work, a machine (AI) may alert someone to underperformance due to leading indicators around productivity.

There are also risks to organizational resilience beyond fairness. Such an overemphasis on automation could erode institutional knowledge and undermine HR professionals' critical thinking skills, which can make businesses susceptible to errors or low-quality results from AI (Powell, 2024). Workers, moreover, may become uninspired, lose moral, and trust the leadership less if they are aware of how generative AI makes decisions about their work (Farinha & Pina, 2025). Organizations could consider adopting "mixed models of decision-making," in which AI plays an advisory but not decision-making role (Wang & Pashmforoosh, 2024) to address these potential harms. In order to sustain the consistency between the efficiencies associated with technology and the values essential to an ethical approach to HRM, the "human in the loop " approach to decision-making ensures that the decision nonetheless remains subject to the judgements made by the humans involved. HR practice's potential to evaluate these suggestions made by technology and to recognize the need for involvement by the humans involved could be supplemented by the provision of AI literacy for HR staff.

2.4 *Legal and Compliance Issues in HR Contexts*

According to Cheong et al. (2024), a complex interplay of labour, anti-discrimination, and data protection regulations applies when generative AI and LLMs are integrated into HR-related activities. Organizations that fail in these respects face a risk of regulatory fines, costly litigation, and damage to their brand. According to guidelines issued by the Equal Employment Opportunity Commission, or EEOC, of the United States, AI tools in hiring violate Title VII of the Civil Rights Act when generating unintended discriminatory results. Similarly, the General Data Protection Regulation, or GDPR, of the EU restricts automated decision-making, which significantly affects individuals and imposes obligations of transparency, data minimization, and the right to human intervention. Legal risks are not confined to data privacy and discrimination issues. AI-powered HR solutions can also create labour law and contract problems, warns (Gerards & Borgesius, 2022). For example, algorithmic scheduling systems will violate employment contracts if the scheduling systems arbitrarily change working hours without adequate notice. Union-negotiated agreements may be violated even in jurisdictions with a strong tradition of collective bargaining.

Additionally, a more important requirement in explainability has emerged. Nowadays, various agencies and regulators are stressing how important it is that businesses provide employees with clear and intelligible explanations of how AI makes decisions that affect them. Arguably, organizations are using AI in a way that minimizes legal risks through various practices, including performing AI compliance audits, hiring legal counsel in tech procurements, and designing transparent AI governance frameworks. Developing distinct lines of accountability and ensuring there are identifiable human managers across various important decisions in organizations, including those in HRM, not only makes legal prescriptions but also models ethical behaviour as well. Though there are chances that AI models, including those in AI-G, would raise a revolutionary wave in managing talent through their capabilities in generative activities, their deployment would need to be established within a strong governance structure that eliminates prejudices, protects privacy, enhances humanistic agency, and adheres to morality and legality.

3 Implications for Africa and Developing Economies

Table 2 summarizes the key implications of adopting generative AI and large language models in HRM across Africa and other developing economies. It also brings to bear how these technologies can help address HR capability gaps in various sectors, in addition to possibilities of digital colonialism and poor localization of technologies in many African countries, outlining how public and private sectors' use of e-HRM systems will differ in other references.

Table 2 Implications of generative AI and LLMs for Africa and developing economies

Theme	Description	Examples	Key references
Potential to close HR capability gaps	AI can automate HR tasks, enhance recruitment efficiency, and provide personalized learning opportunities, addressing shortages of skilled HR professionals and limited training infrastructure.	Fuzu (Kenya) uses AI for job–candidate matching; Eneza education uses AI for personalized learning in East Africa.	World Bank (2021); Mhlanga (2023); Otieno (2022); UNESCO (2022)
Risks of digital colonialism or lack of localized training data	Over-reliance on AI systems developed in the global north can embed foreign norms, marginalize African languages, and store sensitive data abroad, threatening data sovereignty.	AI recruitment tools are undervaluing African university qualifications, and there is a lack of African language datasets in LLMs.	African Union (2020); Couldry & Mejias (2019); Gillwald et al. (2022); Ndhlovu (2021);
Talent Management in Public vs. private sectors	Private sector adopts AI faster due to better funding and infrastructure; public sector adoption lags, risking talent drain from essential public services.	Safaricom (Kenya) uses AI analytics for employee engagement; Rwanda pilots AI in health workforce allocation.	African Development Bank (2021); Deloitte (2023)WHO (2022)

3.1 *Potential to Close HR Capability Gaps*

Large language models and generative AI have the potential to address long-standing gaps in human resource competencies in poor countries, such as those in Africa. Most organizations in these regions suffer from an acute shortage of human resources specialists, limited access to state-of-the-art hiring technologies, and a lack of training facilities. Whichever way one looks at it, resource limitations notwithstanding, organizations can simplify HR processes by embracing AI-enabled tools to automate routine tasks such as performance reviews, interview scheduling, onboarding, and candidate screening. In order to improve recruiting efficiency for SMEs that might not have dedicated HR staff, the Kenyan recruitment portal Fuzu, for instance, matches candidates to jobs based on skills, experience, and behavioural attributes using AI algorithms (Otieno, 2022). Smaller businesses may be better equipped to compete for talent against larger companies with more human resources.

Additionally, generative AI can make learning and growth opportunities more accessible. By using AI to provide individualized learning materials in regional languages, adaptive e-learning platforms such as Eneza Education in East Africa help close skill gaps and improve job preparedness (Mhlanga, 2023). Mobile-based training tools and AI-driven language translation can help nations with sizable rural populations overcome linguistic and geographic obstacles to workforce development (Ndhlovu, 2025). Industries and governments can also use predictive workforce analytics to estimate future labour market demands and pinpoint skills shortages (UNESCO, 2022). According to economic priorities, this enables targeted support for reskilling programs in industries including healthcare, digital services, and renewable energy (Li, 2024). However, to fully reap these benefits, funding for digital infrastructure, reasonably priced internet access, and training for HR professionals in AI literacy are all necessary (Mazurchenko & Marsikova, Mazurchenko and Maršíková 2019; Twala & Ndhlovu, 2024). It may not remain that way without these primary constituent elements, and without them, revolutionized by generative AI and helping to erase skill gaps in HR.

3.2 Risks of Digital Colonialism or Lack of Localized Training Data

Although generative AI has many advantages, there are drawbacks to its use in Africa, most notably the risk of "digital colonialism" and the insufficient localization of training data (Amoah, 2024). A situation known as "digital colonialism" occurs when markets in the Global South are dominated by technology platforms, data infrastructures, and artificial intelligence (AI) systems primarily built in the Global North. This frequently results in integrating foreign cultural norms, values, and priorities into local systems (Couldry & Mejias, 2019). African languages, histories, and cultural settings are not well represented in the datasets used to train the majority of LLMs, which are primarily English-language sources. Thus, characteristics of jobs, training programs, and employee evaluation may not reflect the realities of a particular area, owing to lingual and cultural biases inherent in the outputs of AI (Ndhlovu, 2021). An AI-based hiring tool that was preponderantly trained on Western labour domain data, for example, would not recognize special skills enjoyed by that area, nor recognize qualifications from African colleges. Another significant issue in this regard relates to data sovereignty. Multinational companies often operate servers in other countries that hold vital information regarding Africa-based workers who were assessed by an AI-based HR tool. It raises suspicions regarding long-term data superiority in relevant labour domain information and data privacy regulations (Gillwald et al., 2022). For example, politicians in Africa have cautioned that the ability of the country to implement data privacy regulations may not succeed owing to overdependence on foreign-owned cloud services.

For strategic interventions to address these dangers, there is a pressing need to make strategic investments in regionally developed AI models, to develop regional AI datasets, and to build African data infrastructure. In order to guarantee that regionally developed AI systems reflect African economic and social interests rather than those based on suppositions obtained elsewhere, joint efforts such as those set up through initiatives such as the African Union's Digital Transformation Strategy (2020) aim to develop indigenous technical capabilities.

3.3 Talent Management in Public Vs. Private Sectors

In the case of Africa and other emerging economies worldwide, the private and public sectors in various HRMs show different patterns in the adoption of generative AI technology (Budhwar et al., 2023). In the private sectors of these economies, for example, the adoption of AI has been quicker compared to the public sectors over time (Llorens, 2021). In the private sectors of emerging economies like Africa and Asia, the adoption has mainly taken place in large organizations and local businesses like the Kenyan telecommunication firm Safricom; the firm has managed to take advantage of the strong information technology base and financial status to improve internal dynamics and increase employee engagement through the adoption of AI technology for analytics (Deloitte, 2023). On the other side, owing to outmoded processes, bureaucratic decisions, and budget constraints, the adoption process is often slow in the public sector (African Development Bank, 2021). Nevertheless, there are significant opportunities for an efficient workforce with regards to the use of artificial intelligence in the public sector, particularly in key areas such as administration, healthcare, and education (Ajayi et al., 2024).

To exemplify, artificial intelligence is reportedly analysed as a potential aspect that can be used to enhance the delivery of healthcare outfits by linking medical practitioners with underserviced areas, particularly Rwanda (WHO, 2022). In case the adoption process continues to be slow with regards to artificial intelligence, there could be an enormous brain drain from the public to the private sector owing to its advanced use of artificial intelligence technology to deliver better opportunities through lifestyle and pay packages that are compatible with a work environment. There could also be a skill drain with regards to key fields such as healthcare and education. Thus, some policies that could help rectify this mismatch include government incentives for AI adoption in public institutions, technology-sharing programs between sectors, and AI capacity-building programs for public sector HR teams. In this vein, ensuring a more equitable distribution of benefits from AI-driven HR transformation across both sectors will be fundamental to attaining inclusive economic growth and societal resilience in developing economies.

4 Future Trends and Strategic Recommendations

4.1 Responsible AI Use in HR

The responsible application of AI in HR is increasingly being outlined by international standards, industry-specific guidelines, and risk-based governance. A risk-based approach that emphasizes context, impact, and suitable controls throughout the AI lifecycle is supported by the U.S. National Institute of Standards and Technology (NIST). The NIST AI Risk Management Framework (AI RMF 1.0) and its Generative AI Profile translate abstract ideas into real-world processes like post-deployment monitoring, secure operations, and bias testing. These procedures are particularly important in HR settings where discrimination, privacy, and explainability are significant concerns. International policy frameworks are also emerging. By reiterating these commitments in the Organization for Economic Co-operation and Development's (OECD) updated AI Principles, HR directors can integrate fairness, openness, robustness, and accountability into vendor contracts and internal governance processes (Powell, 2024). Standardization projects like ISO/IEC 42001 provide the requirements for an Artificial Intelligence Management System (AIMS) that may integrate AI ethics, risk, and quality management into existing organizational systems. Regulations are further influencing HR practices. The European Union's AI Act (European Parliament, 2016) designates employment-related AI systems as "high-risk" and establishes standards for data governance, transparency, and human oversight. This means that AI-driven decisions must have documented testing, explainability, and explicit human approval for HR teams.

To achieve 'responsible by design' adoption, HR departments should strategically institutionalize AI governance through formal policies and risk registers, mandate that vendors show that they comply with ISO/IEC 42001, perform bias and privacy impact assessments before and after deployment, and provide employees with plain-language disclosures explaining how AI is used in decision-making.

4.2 Human-in-the-Loop Hybrid HR Models

In HR decision-making, the "human-in-the-loop" method is becoming a recommended practice to balance AI effectiveness with human judgement (Sidra & Wagan, 2026). This model takes into account two realities: even with proper calibration, AI systems may drift or perform poorly for specific subpopulations, and hiring decisions often require due process, empathy, and contextual knowledge (Lazaros et al., 2026). This concept has been increasingly entrenched in regulatory frameworks. For instance, the EU AI Act requires that any high-risk AI system should have meaningful human oversight, providing the ability to intervene and alter outputs (Laux, 2024). Relatedly, it's been made clear by the Equal Employment Opportunity Commission ("EEOC") within the United States "that human review is a key function because employers remain liable for any concerns about discrimination even if they have used a third-party AI tool." It must also be acknowledged that effective hybrid HR solutions possess a multitude of features, which include "two-stage decision-making using AI as a first screen and human decision-making as a second screen," "documentation and explainability features such as data sheets and model cards created specifically to meet the needs and requirements of the HR function," and "appeal and review features to allow candidates and employees to review and increase the overall accuracy and fairness through human reviewer feedback used to retune AI models." By doing this, organizations can successfully meet diverse legislative requirements, increase the trust from staff, and ensure human interaction is maintained in any highly automated process through the application and integration of HITL solutions.

4.3 Policies for Transparent AI Use in People Management

In the ever-evolving domain of reliable AI in human resource departments, transparency is also becoming an important feature. Clear documents, disclosures, and traceability are needed in AI systems affecting human populations under the EU AI Act and the OECD AI Principles (European Parliament, 2016). Implementing transparency in the human resource sector involves displaying easily readable documents to employees and job seekers to explain the use of AI technologies, the data used by the technologies, and the employee/seeker rights, such as the human review right. For operational transparency, as proposed by the ISO/IEC, it is also important to maintain detailed records of the inputs, outputs, overrides used, and the origins of data used to train the AI technologies and any constraints used. ISO/IEC 42001 is a framework which offers detailed guidance on how to integrate such schemes into an auditable management system. Such review and restitution mechanisms need to be integrated into transparency strategies. In addition, the requirement to avoid harm is achieved by the implementation of tests prior to deployment, the need to continually monitor for bias, and the establishment of rollback conditions. The legal requirement to ensure explainability and easy access to appeal is underscored in the latest legal actions taken by the EEOC, including those associated with the bias in the use of the AI recruitment strategies offered by the vendor. Vendor management is arguably as critical because it requires the disclosure of evaluation documents to comply with the guidelines set out and to collaborate in audits. Senior-level assurance is required to align with the ethical considerations with regards to people management. An action here is the Chief Human Resources Officer and the AI Governance lead.

5 Conclusion

The Fourth Industrial Revolution has been followed by a number of innovative tools being implemented by different industries to ensure cost-effectiveness and efficacy within businesses. Among the different departments such as HRM where different innovative tools such as generative AI and LLMs can be implemented were identified in the present chapter, followed by an elaboration of the complexities involved in integrating generative AI and LLMs within an HRM function of a business organization, highlighting different benefits of integrating generative AI and LLMs within an HRM function of a business organization. The LLM can be used to create job descriptions that are more captivating, precise, and diversified. The LLM can also be used to efficiently sift through candidates by not only focusing on matching the keywords of resumes, but rather trying to understand the context of the resume presented by an applicant. Further, LLMs can also be used to automate further correspondence with each candidate, for instance, setting an interview date or sending a rejection letter to a candidate. Such LLMs can be used to generate customized questions to be asked during an interview based on each candidate's background or particular skills to be tested for a particular position. It has been identified that whereas literature on the advantages of these technologies has been concerns-laden, challenges associated with it are not well-explored. Thus, the chapter noted a number of challenges in incorporating LLMs and generative AI into HRM. These include an excessive reliance on automation rather than human resources, privacy and surveillance issues, and biases in training data and outputs. Therefore, it is concluded that human resources should be complemented rather than purposefully replaced when integrating new digital tools into HRM. Businesses must make investments to enable their employees to use cutting-edge digital tools to enhance operations.

References

African Development Bank. (2021). *African economic outlook 2021: From debt resolution to growth*. African Development Bank Group.

African Union. (2020). The digital transformation strategy for Africa (2020–2030). African Union Commission.

Ajayi, A. J., Agbede, O. O., Akhigbe, E. E., & Egbuhuzor, N. S. (2024). Enhancing public sector productivity with AI-powered SaaS in e-governance systems. *International Journal of Multidisciplinary Research and Growth Evaluation, 5*(1), 1243–1259.

Amoah, L. G. A. (Ed.). (2024). *Examining the rapid advance of digital technology in Africa*. IGI Global.

April, K., & Daya, P. (2025). The use of AI in HRM and management processes: The promise of diversity, equity, and inclusion. In *AI and diversity in a Datafied world of work: Will the future of work be inclusive?* (Vol. 12, pp. 97–123). Emerald Publishing Limited.

Ball, K. (2010). Workplace surveillance: An overview. *Labor History, 51*(1), 87–106. https://doi.org/10.1080/00236561003654776

Ball, K. (2021). *Electronic monitoring and surveillance in the workplace*. European Commission Joint Research Centre.

Bankins, S. (2021). The ethical use of artificial intelligence in human resource management: A decision-making framework. *Ethics and Information Technology, 23*(4), 841–854.

Barocas, S., Hardt, M., & Narayanan, A. (2023). *Fairness and machine learning: Limitations and opportunities*. MIT Press.

Bhalla, N. K. R., & Suribabu, A. K. G. P. (2024). Ethical implications of AI adoption in HRM: Balancing automation with human values. *Journal of Informatics Education and Research, 4*(3). https://doi.org/10.52783/jier.v4i3.1337

Bryson, J. J. (2018). Patiency is not a virtue: AI and the design of ethical systems. *Ethics and Information Technology, 20*(1), 15–26. https://doi.org/10.1007/s10676-018-9448-6

Budhwar, P., Chowdhury, S., Wood, G., et al. (2023). Human resource management in the age of generative artificial intelligence: Perspectives and research directions on ChatGPT. *Human Resource Management Journal, 33*(3), 606–659.

Chen, Z. (2023). Ethics and discrimination in artificial intelligence-enabled recruitment practices. *Humanities and social sciences communications, 10*(1), 1–12.

Cheong, I., Caliskan, A., & Kohno, T. (2024). Safeguarding human values: Rethinking US law for generative AI's societal impacts. *AI and Ethics*, 1–27.

Couldry, N., & Mejias, U. A. (2019). *The costs of connection: How data is colonizing humanlife and appropriating it for capitalism*. Stanford University Press.

Dastin, J. (2018, October 10). Amazon scraps secret AI recruiting tool that showed bias against women.. Retrieved July 17, 2025, from https://www.reuters.com/article/us-amazon-com-jobs-automation-insight-idUSKCN1MK08G

De Almeida, P. G. R., Dos Santos, C. D., & Farias, J. S. (2021). Artificial intelligence regulation: A framework for governance. *Ethics and Information Technology, 23*(3), 505–525.

Deloitte. (2023). *2023 global human capital trends: Leading in the boundaryless world*. Deloitte Insights.

Du, J. (2024). Exploring gender bias and algorithm transparency: Ethical considerations of AI in HRM. *Journal of Theory and Practice of Management Science, 4*(03), 36–43. https://doi.org/10.53469/jtpms.2024.04(03).06

Ebert, I., Wildhaber, I., & Adams-Prassl, J. (2021). Big data in the workplace: Privacy due diligence as a human rights-based approach to employee privacy protection. *Big Data & Society, 8*(1). https://doi.org/10.1177/20539517211013051

Farinha, J., & Pina, M. D. F. A. (2025). The impact of positive leadership on employee engagement, productivity, and talent retention in the era of artificial intelligence. In *Approaching employee experience management with data science* (pp. 197–222). IGI Global Scientific Publishing.

Floridi, L., Cowls, J., Beltrametti, M., et al. (2018). AI4People—An ethical framework for a good AI society: Opportunities, risks, principles, and recommendations. *Minds and Machines, 28*(4), 689–707. https://doi.org/10.1007/s11023-018-9482-5

Gerards, J., & Borgesius, F. Z. (2022). Protected grounds and the system of non-discrimination law in the context of algorithmic decision-making and artificial intelligence. *Colo. Tech. LJ, 20*, 1.

Gillwald, A., Moyo, M., & Rademan, B. (2022). Data governance in Africa: Ensuring equity and inclusion in a digital economy. *The African Journal of Information and Communication, 29*, 1–17. https://doi.org/10.23962/10539/32153

Lancieri, F. (2022). Narrowing data protection's enforcement gap. *Maine Law Review, 74*, 15.

Laux, J. (2024). Institutionalised distrust and human oversight of artificial intelligence: Towards a democratic design of AI governance under the European Union AI Act. *AI & Society, 39*(6), 2853–2866.

Lazaros, K., Vrahatis, A. G., & Kotsiantis, S. (2026). Human-in-the-loop artificial intelligence: A systematic review of concepts, methods, and applications. *Entropy, 28*(4), 377.

Li, L. (2024). Reskilling and upskilling the future-ready workforce for industry 4.0 and beyond. *Information Systems Frontiers, 26*(5), 1697–1712.

Llorens, J. J. (2021). Rapid advances in HRM technologies and public employment systems: A research agenda for acquiring and managing talent. *Handbook of Public Administration*, 272–281.

Mazurchenko, A., & Maršíková, K. (2019). Digitally-powered human resource management: Skills and roles in the digital era. *Acta Informatica Pragensia, 8*(2), 72–87.

Mehrabi, N., Morstatter, F., Saxena, N., et al. (2021). A survey on bias and fairness in machine learning. *ACM Computing Surveys, 54*(6), 1–35. https://doi.org/10.1145/3457607

Mhlanga, D. (2023). Artificial intelligence in education: A panacea for access, equity, and quality in Africa? *Education and Information Technologies, 28*(4), 4511–4532. https://doi.org/10.1007/s10639-023-11650-0

Ndhlovu, E. (2025). Artificial intelligence and data science in African food systems. *IAHRW International Journal of Social Sciences Review, 13*(3), 556–563.

Ndhlovu, F. (2021). Decolonising language education in Africa: Lessons from multilingual contexts. *Current Issues in Language Planning, 22*(4), 365–384. https://doi.org/10.1080/14664208.2020.1853142

Onyekwelu, N. P., Ezeafulukwe, C., Owolabi, O. R., et al. (2024). Ethics and corporate social responsibility in HR: A comprehensive review of policies and practices. *International Journal of Science and Research Archive, 11*(1), 1294–1303.

Otieno, J. (2022). May 18. AI recruitment platforms transform hiring in Africa. https://www.businessdailyafrica.com

Parliament, E. (2016). *Regulation (EU) 2016/679 of the European Parliament and of theCouncil* (general data protection regulation). *Official Journal of the European Union.*

Porkodi, S., & Cedro, T. L. (2025). The ethical role of generative artificial intelligence in modern HR decision-making: A systematic literature review. *European Journal of Business and Management Research, 10*(1), 44–55.

Powell, J. R. (2024). Human resource professionals' perceptions of Trust in Explainable Artificial Intelligence Hiring Software (Doctoral dissertation, National University).

Raghavan, M., Barocas, S., Kleinberg, J., & Levy, K. (2020). Mitigating bias in algorithmic hiring: Evaluating claims and practices. *Proceedings of the 2020* Conference on fairness, Accountability, and Transparency, 469–481. doi:https://doi.org/10.1145/3351095.3372828.

Ruokonen, M. (2025). Applying AI in key European industries—Strengthening competitiveness and innovation in Europe. Sitra Working Papers. https://www.sitra.fi/wp/wp-content/uploads/2025/03/sitra-applying-ai-in-key-european-industries.pdf.

Segarra-Blasco, A., Tomàs-Porres, J., & Teruel, M. (2025). AI, robots and innovation in European SMEs. *Small Business Economics, 65*, 719–745. https://doi.org/10.1007/s11187-025-01017-2

Shrestha, Y. R., Ben-Menahem, S. M., & von Krogh, G. (2019). Organisational decision-making structures in the age of artificial intelligence. *California Management Review, 61*(4), 66–83. https://doi.org/10.1177/0008125619862257

Sidra, S., & Wagan, S.M. (2026). Human-in-the-loop: How managers can govern AI decisions in HR practices. *Journal of Modelling in Management*, 1–23.

Spinello, R. A. (2021). Corporate data breaches: A moral and legal analysis. *Journal of Information Ethics, 30*(1), 12–32.

Stahl, B. C., & Wright, D. (2018). Ethics and privacy in AI and big data: Implementing responsible research and innovation. *IEEE Security & Privacy, 16*(3), 26–33. https://doi.org/10.1109/msp.2018.2701164

Twala, K., & Ndhlovu, E. (2024). Artificial intelligence and robotic adoption by customers in the global hotel sector. In E. In Ndhlovu, K. Dube, & T. Makuyana (Eds.), *Tourism and hospitality for sustainable development - volume three: Implications for customers and employees of tourism businesses*. Springer International.

UNESCO. (2022). *Reimagining our futures together: A new social contract for education.* UNESCO Publishing.

Wang, J., & Pashmforoosh, R. (2024). A new framework for ethical artificial intelligence: Keeping HRD in the loop. *Human Resource Development International, 27*(3), 428–451.

WHO. (2022). *Rwanda's digital health transformation.* World Health Organisation Regional Office for Africa.

Worek, M., & Aaltonen, P. (2025). 'AI adoption challenges in family-owned firms: A case study.' In: Aaltonen, P., Kurvinen, E. (eds) Contemporary issues in industry 5.0. *Technology, Work and Globalisation.* Palgrave Macmillan,. doi:https://doi.org/10.1007/978-3-031-74779-3_9.
World Bank. (2021). *World development report 2021: Data for better lives.* World Bank Group.
Zuboff, S. (2019). *The age of surveillance capitalism: The fight for a human future at the new frontier of power.* PublicAffairs.

Brian Mabuyana (PhD) earned his PhD in Business Management from the University of the Free State, Bloemfontein. He is a Post-Doctoral Fellow in the Department of Marketing and Retail Management at the University of South Africa (UNISA). He is a digital marketing, fintech, tourism, entrepreneurship, sustainability, artificial intelligence, and consumer behavior researcher. He has published articles in reputable international journals such as Sustainable Technology & Entrepreneurship (Elsevier), Young Consumers (Emerald), Cogent Business & Management (Taylor & Francis), Business Analyst Journal (Emerald), Journal of Economic Criminology (Elsevier), Benchmarking: An International Journal (Emerald), Management & Sustainability: An Arab Review (Emerald), African Journal of Innovation and Entrepreneurship (Sabinet), Strategic Business Research (Elsevier) among others. Some of his papers have been presented at a local conference. In addition, he has supervised honors students and master's students. He is currently supervising PhD students.

David Mhlanga is an award-winning Economist, Researcher, and Higher Education Professional with extensive experience in teaching, research, and policy consulting across Africa and Australia. He currently serves as Extraordinary Professor in the Department of Financial Governance, College of Accounting Sciences at the University of South Africa (UNISA), and as an Extraordinary Associate Professor in the Department of Economics at North-West University (South Africa). He is also a Teaching Associate in the Department of Economics at Monash University (Australia) and a Lecturer at ACAP University College (Australia). Professor Mhlanga is the Principal Consultant at the Econowrite Institute for Policy Innovation & Development (EIPID), based in Melbourne, where he leads interdisciplinary research and policy initiatives focusing on digital transformation, financial technology (FinTech), artificial intelligence (AI), and sustainable development.

Mufaro Dzingirai is a Senior Lecturer at Harold Pupkewitz Graduate School of Business, Namibia University of Science and Technology, Namibia and Research Fellow at INTI International University, Malaysia, as well as a Consultant at the Econowrite Institute for Policy Innovation & Development (EIPID), Australia.

Emmanuel Ndhlovu is a Senior Research Fellow at the University of Johannesburg, South Africa. He holds a PhD in Development Studies from the University of South Africa. His research interests cover a wide range of topics, including agrarian political economy, food systems, food security, land reform, agriculture and agribusiness, and social policy. His recently authored books include Agriculture, Autonomous Development and Prospects for Industrialization in Africa and Social Policy in Southern Africa: Prospects and Possibilities. He has edited more than 10 books.

Human Resource Analytics: Challenges and Prospects for Human Resource Effectiveness

Ibrahim Olanrewaju Lawal

Abstract Although Human Resource Analytics (HRA) is increasingly used and the question of how it can influence employee productivity remains unanswered. Human Resource Analytics is an enhancement of Human Resources practice that uses data analysis to advance HR practices, improve employee engagement, performance, and decision-making, and enhance organizational growth, efficiency, and strategic workforce planning. The current research aims to fill this gap by investigating the impact of HRA on job satisfaction, engagement, performance, and other aspects of the workplace. The study adopted the Systematic Reviews and Meta-analyses (PRISMA) to achieve smooth, superior, and improved comprehension of the examination of the gathered information and analysis. The results indicate that HRA is positively related to employee satisfaction, engagement, and overall performance, thereby enhancing HR's capacity to address workforce needs. Nevertheless, issues of low data quality, lack of analytical skills, and ethical concerns, especially regarding privacy and fairness, affect the effectiveness and credibility of HRA initiatives. To get the best out of HRA, the chapter suggests that organizations adopt effective data management systems, improve HR professionals' analytical capabilities through training, and develop ethical standards. Such measures will contribute to transparency, employee confidence, and the successful implementation of data-oriented HR practices.

Keywords Analytical skills · Data quality · Employee job satisfaction · Ethical concerns · Performance

1 Introduction

Some of the new technologies that organizations are currently using include Data analytics, Artificial Intelligence (AI), Robotic Process Automation (RPA), cloud computing, the Internet of Things (IoT), and blockchain (Madhani, 2023). The rapid

I. O. Lawal (✉)
Department of Business Management, GBS/University of Suffolk, Birmingham Campus, United Kingdom
e-mail: ilawal@globalbanking.ac.uk

M. Dzingirai and S. O. Atiku (eds.), *Strategic Human Resource Analytics and Intelligent Data-Driven Systems*, https://doi.org/10.1007/978-981-92-2623-8_11

pace of these Information Technologies (IT) has also driven a digital revolution, as companies are using IT to collect, store, process, access, and share HR information to leverage previously unexplored opportunities. The Human Resources HR departments are also moving towards the use of analytical methods to enhance employee performance. The application of data analytics to assess HR functions is called Human Resource Analytics (HRA), which holds great potential to enhance how organizations manage their workforce, improve productivity, and increase employee satisfaction (Dunmade et al., 2023; Madhani, 2022). With advanced tools to interpret data in relation to the workforce, HR managers will be able to make evidence-based decisions that will help enhance hiring, retention, engagement, and performance, while also tackling complex challenges (Kumar et al., 2022). Despite its advantages, HRA implementation poses challenges, including data access, privacy, and the need to develop skills within HR departments. The development of data analytics in HR is part of the broader drive towards technology and organizational strategy. HRA represents a shift from past HR practices based on intuition and experience to more scientific practices that use data to understand how employees behave and predict future trends. Nowadays, HRA is applied to more than 70 percent of organizations worldwide to understand how employees engage, perform, and stay with organizations (Deloitte, 2023). It implies that HR specialists can use data-driven methods to identify the key drivers of employee satisfaction and take significant steps to boost morale and productivity.

HRA is divided into descriptive, predictive, and prescriptive analytics. Descriptive analytics relies on historical data and trends; predictive analytics uses statistical models to predict future outcomes; and prescriptive analytics involves adopting actionable strategies based on results (Marler & Boudreau, 2017). All these types of analysis help HR managers make informed decisions, minimize turnover, and create a productive work environment. For example, through predictive analytics, HR departments can identify employees likely to exit, and retention strategies can be tailored accordingly (Zebua et al., 2024).

Employee performance, including job satisfaction, engagement, performance, and retention, is a significant indicator of an organization's health. Research has shown that when organizations focus on improving these results, their profitability and productivity increase. To illustrate, a Gallup (2022) report found that productivity can increase by 23 percent when employees are highly engaged and turnover is lower than in companies with disengaged workforces. Accordingly, the HRA's capacity to measure and optimize employee performance is offering business organizations a competitive advantage by underscoring the importance of HR in strategic decision-making (Kumar et al., 2022).

Despite the advantages, there are no challenges associated with implementing HRA. One of the most significant problems is data accessibility. HR analytics-relevant data, including performance indicators, employee engagement indices, and turnover statistics, are likely to be stored in separate systems that form data silos, impeding functional analysis (Rasmussen & Ulrich, 2015). To make HRA effective, organizations should develop integrated systems that centralize and streamline data collection. In addition, the data must be accurate, as inaccurate data may lead to incorrect conclusions that adversely affect employee morale and efficiency.

Ethical issues and data privacy are also serious challenges. HR data contains sensitive information that frequently raises concerns about the methods of data collection, storage, and analysis. Laws such as the General Data Protection Regulation in the European Union (EU) highlight the need to use employee data responsibly (European Union, 2018). HR managers should have adequate data protection policies in place to facilitate compliance and build trust among employees, as any misappropriation of data could lead to negative publicity for the organization and attract legal repercussions. The other major issue in implementing HRA is the skills gap in HR departments. The problem with analytics tools is that, in many cases, HR professionals lack the technical skills to analyse data effectively (Rasmussen & Ulrich, 2015). To resolve this, organizations must invest in HR upskilling, train staff in data analytics, and introduce them to cutting-edge data analysis software. According to a Deloitte (2023) study, half of HR professionals reported they cannot make decisions based on data due to insufficient analytics training.

These challenges must be addressed through a strategic plan that includes technology, employee training, and data governance. To start, organizations should incorporate HRA tools into their HR systems to avoid data silos and ensure a smooth flow of information. Cloud-based Human Capital Management (HCM) systems have gained popularity because they enable HR professionals to access data in real time, enabling more rapid and precise analysis (Rana & Sharma, 2023). These systems also enable the integration of HR and IT and better alignment of HRA with the overall organizational objectives. The creation of a data-driven HR culture relies on improving employee education and upskilling. The gap in analytics skills can be addressed by organizations providing specific training programmes and supporting collaboration with data science departments. This will ensure that HR professionals are not only aware of how to use analytics tools but also able to interpret the data to make informed choices. This ability of the HR department to build strategic capacity transforms HR from a supporting unit to a key determinant of business success.

Data governance ethics are fundamental. HR managers must focus on transparency, ensuring employees understand how their data will be used, and taking precautionary measures to prevent sensitive information leaks. This fosters confidence in the organization, facilitating easier data collection and improving data accuracy (European Union, 2018). The ethics must also be applied to the analytics process, ensuring that algorithms and predictive models used in HRA do not reinforce biases or inequalities. By integrating ethical behaviours into HRA, organizations can create a respectful and equal working environment (Wang et al., 2024).

With organizations' investment in HRA, the role of HR is going to transform into a more strategic one. The new possibilities of HRA are enabled by emerging technologies, such as machine learning (ML) and artificial intelligence (AI), which can automate data analysis and improve predictive capabilities. For example, AI-powered HRA tools can assess an employee's engagement and make immediate suggestions to improve their working environment (Forbes, 2023). Nevertheless, as technology advances, the need to manage data responsibly and consider ethical issues will remain essential.

HRA is a growing phenomenon, but significant knowledge gaps remain. Although much has been said about the literature, what has not been clearly stated is how HRA affects key employee outcomes, especially in multi-dimensional organizational settings. In addition, the majority of the literature focuses on the potential benefits of HRA while ignoring other long-term problems, such as data quality, a lack of analytical skills, and ethical concerns related to data privacy and fairness (Deloitte, 2023; Wang et al., 2024). The proposed research addresses such gaps by conducting a systematic review of recent empirical data using the PRISMA framework to determine the extent to which HRA promotes the effectiveness and productivity of organizations and their employees. In particular, it is intended to delve into whether evidence-based HR interventions can improve performance, engagement, and satisfaction, and in which scenarios they should be implemented to be effective.

2 Objectives of the Study

- To determine the influence of human resource analytics on job satisfaction, engagement, and performance.
- To examine how human resource analytics influences data management systems and enhances HR professionals' analytical skills.

3 Methodology

The present study applied Systematic Reviews and Meta-analyses (PRISMA) to promote transparency, completeness, and quality of reporting. It is a seamless process that guarantees high-quality and trustworthy systematic review analysis (Haddaway et al., 2022), and it has continued to attract academic scholars' interest in the field of management (Jimoh, 2025). The use of PRISMA illustrates the systematic screening process undertaken in a literature review on the topic "Human Resource Analytics: Challenges and Prospects for Human Resource Management". The process began with 50 records identified from databases, after removing six duplicates and one automation-excluded record. The 43 records were retrieved and searched for full-text assessment, and they were assessed for eligibility accordingly. Of the 43 full-text reports that were assessed, five were excluded due to the following: they were not relevant to the research purpose, some had poor data, some were duplicate entries of others already included, some were conference abstracts, and some were non-English language records.

There were other duplicate studies or conference abstracts, which were excluded by the review protocol since they were not methodologically elaborate. No such studies were found in other sources, such as websites, organizational publications, or citation-tracking, despite browsing 479 citations and two websites/organizational sources. This suggests a reasonably narrow base of stringent, peer-reviewed research on subjects that points towards not high-quality empirical work in the area that is readily available. Finally, seven more studies were added, bringing the total to 45. The smaller number of new eligible studies confirms that, while there is growing interest in human resource analytics, there remain only a few methodologically sound, data-rich publications. This supports the argument that one of the significant issues in the field is not merely the underleveraging of the Analytics platform in HRM, but also a dearth of rigorous, evidence-based research that meets systematic review standards. The limited coverage of studies emphasized the need for further empirical research on the strategic importance and operational hurdles of using analytics in HR decision-making, particularly in activities such as people management, employee planning, and performance optimization.

4 Results

4.1 The Concept of Human Resource Analytics

In today's intricate corporate environment, every choice is supported by data and facts (Bassi, 2022). Every function in HR, from hiring to separation, depends on data. Previously, HR was thought of as a theoretical field. According to Moore et al. (2011), HRA is crucial for accurately interpreting this data, which makes it helpful. The act of gathering and evaluating HR-related data to improve an organization's workforce is called HR analytics (Falletta & Combs, 2021; Margherita, 2022). Other names for the interaction include labour analytics, individual investigation, and ability evaluation (Werkhoven, 2019). Methods of data exploration are employed to link HR composed data to organizational and HR domains (Chatterjee et al., 2022).

HR analytics provides quantifiable proof of how HR creativity supports organizational objectives and tactics (Qamar & Samad, 2022). According to Chalutz Ben-Gal (2019), HR analytics provides a data-supported understanding of what has been done and what is in place, enabling businesses to progress and improve their future strategies. Margherita (2022) proposes four phases as the steps involved in HR analytics. To gain the problem-solving insights HRA promises, data collection is the first step. The second step is to compare the data to other data, such as standards, midpoints, or verifiable data, and estimate the results. The analytical stage involves analysing the results. Using insight to inform organizational choices is the last phase.

The practical application of HRA enhances employee performance by helping them identify their strengths and development areas. An example is predictive analytics, which may help the HR group foresee performance issues and offer solutions before they occur, including targeted training (Gartner, 2023). HRA can be used to make performance appraisals more relevant, as it allows following up on current data across various metrics and fostering a culture of continuous improvement.

According to McKinsey (2023), employees who receive consistent, data-driven, positive feedback tend to perform exceptionally well, as it aligns their personal targets with organizational outputs.

The insights produced by HRA have a significant effect on the retention rates. By examining employee turnover data, HR teams can identify trends and factors that drive disengagement or dissatisfaction (Dhankhar & Singh, 2022). HR managers ought to do this by initiating retention strategies, such as improved work–life balance packages or career development, when predictive models indicate employees are likely to leave. According to a SHRM (2023) study, turnover rates decrease in organizations that predict and reduce turnover using HRA tools, as interventions enhance employee loyalty and commitment.

HRA will also help in employee career development. Organizations could use HRA tools to address skill deficits by designing individualized training programmes that align with each individual's development requirements. The Top Employers Institute (2023) argues that when employees and employers align their career development, job satisfaction will increase and retention will be higher. When employees feel encouraged in their professional development, they are well adjusted to remain with the organization over time. To conclude, HRA plays a key role in achieving positive employee outcomes and high job satisfaction, engagement, performance, retention, and career development, all of which are success factors for the organization.

HRA also has a significant impact on job satisfaction and employee engagement, as it uses data-driven findings to resolve workforce issues. HRA helps organizations to reveal the most important drivers of engagement by analysing employee surveys and feedback, as well as performance indicators. Such a specific action will assist organizations in developing strategies to improve satisfaction and align employees with organizational objectives, thereby increasing productivity and morale (Biswas, 2023; CentraHub CRM, 2023). HRA also contributes to individual employee development by categorizing skills gaps and customizing training programmes. Such a fit between employee desires and professional development will lead to higher levels of satisfaction and retention (Top Employers Institute, 2023). HRA tools also generate predictive analytics that enable the HR team to proactively respond to the risk of disengagement or turnover by providing timely interventions (Avrahami et al., 2022).

HRA-supported real-time feedback systems complement the workplace communication, recognition, and continuous improvement. Employees also feel appreciated, making them even more committed and engaged (CentraHub CRM, 2023). In a broader sense, HRA helps maximize employees' experience by eliminating inefficiencies and balancing work and life, thereby increasing job satisfaction further (Top Employers Institute, 2023). To be effectively implemented, an organizational culture that accepts data-based decision-making and invests in relevant analytical tools and cross-functional knowledge will be necessary. Therefore, HRA is becoming a critical tool for increasing employee engagement, satisfaction, and an organization's success.

HRA is essential for improving employee performance by providing evidence-based information on workforce dynamics. HRA enables organizations to identify performance gaps and introduce personalized interventions by analysing performance metrics, sentiment surveys, and engagement data. To illustrate, predictive analytics can anticipate potential areas of performance concerns and inform personalized training and development programmes to enhance productivity (Avrahami et al., 2022). HRA also enables consistent performance tracking and provides real-time feedback for employees and managers. This is an iterative strategy that ensures individual goals and organizational objectives are aligned and, in turn, enables sustained improvement (McKinsey, 2023). Through such insights, organizations can maximize workforce efforts and drive the overall success of the business.

Human Resource Analytics helps ensure fair, open-minded assessment procedures, eliminating bias and increasing organizational trust. This will introduce a culture of fairness and accountability, which is vital in fostering high performance. The effective use of HRA enhances decision-making quality, and such HR teams are prepared to allocate resources effectively and focus on strategic projects that increase employee productivity (SHRM, 2023). The challenges organizations face in achieving successful integration include data quality, technology adoption, and gaps in analytical skills. Training HR professionals in the use of HR analytics and data governance is a critical measure to maximize the potential of HRA in improving employee performance (Zebua et al., 2024). By integrating the best analytics and strategic HR practices, organizations can foster a culture of continuous improvement and individual performance aligned with corporate objectives.

4.2 Human Resource Analytics and Engagement Level

Data analytics provides an effective way to better understand employees and their level of engagement (Kumar, 2020). It can change the time and area of the HR function (Kumar, 2020). It could be used to manage attrition, define performance criteria, enhance employee relationships, and reduce prejudice during hiring (Kumar, 2020). Most organizations assume that analytics are essential and that they influence the way an organization manages its operations (Kumar, 2020). One of the HR managers at Deloitte asserts that HR branches are evaluating the effectiveness of the right analytics to understand risks and make timely decisions supported by data collected over the years (Kumar, 2020).

Businesses find it challenging to select the best tool when starting analytics projects (Kumar, 2020). Choosing the best tools is typically a classic case of analysis paralysis. As with any other application of information science, the key to HR analytics is how you approach the issue. The algorithms require as much context as possible to compute. When building a model to predict staff attrition, for instance, start with all available factors. The next step is to appropriately reduce the list by examining the relationships and relevance. Data must be contextualized, categorized, calculated, and condensed to produce information (Davenport & Prusak, 2000; Juneja, 2020).

According to Kumar (2020), HR analytics assist with various HR tasks by automating and gradually increasing their cost-effectiveness. In many functional domains and at many levels of management, especially in the Human Resource Department, data exchange is essential (Kumar, 2020). According to Juneja (2020), management degrees can be divided into three main categories: upper, middle, and lower levels. Complete data sharing is available to users at different levels, which are often categorized as executive, middle management, and operations (Juneja, 2020).

Employees are the greatest asset of an organization and the driving force behind achieving a competitive edge in the current unstable global market, yet managing individuals with diverse competencies and coordinating their outputs with the organization's strategy is extremely challenging (Mohammed, 2020). Making decisions requires the creation, analysis, and storage of enormous amounts of data (Mohammed, 2020). To identify top performers from an extensive personnel database, human resource management requires technology that gives managers insights into patterns that emerge across a range of HR operations (Jimoh et al., 2025; Mohammed, 2020).

The Analytics Process for Human Resources analysts use a range of data warehousing techniques, business intelligence tools, and data analysis models to collect information that will be useful for the company's future (Kumar, 2020). By enabling the development of reports, dashboards, and visualization tools, data analytics technology can offer a deeper understanding of HR operations and activities. These include gathering, cleaning, and analysing data, as well as evaluating, implementing, and simplifying plans (Kumar, 2020).

4.2.1 Steps Involved in Human Resource Analytics

Monitoring/Predictive Metrics: Indicators and Key Performance Indicators (KPIs) for predictive Human Resources (HR) systems are already challenging to choose. Future values or events are described as a set of problems, referred to as features, that the machine learning model uses to produce output predictions (AltexSoft, 2019). To develop a model capable of predicting outcomes, the experts should select the most important ones to train it.

One will then be required to consider the list of factors that may lead to a definite future condition, whether good or bad (AltexSoft, 2019). Some sources of data collection include social media, skill assessment systems, recruitment sites, talent management gateways, and the data transformation phase.

Employee data tends to be scattered across paper files, Excel sheets, and multiple HR systems. Accessing data across numerous systems requires significant time and effort (AltexSoft, 2019). There must be one source of information (a centralized data repository) and reliability of data. After gathering all the information about the employees, you will be able to formulate key performance indicators to better understand the significant factors that affect the business targets (AltexSoft, 2019). Data integration technologies transfer data from several systems to the warehouse.

A method of performing this operation is called "ETL (extract, transform, load)" (AltexSoft, 2019). The ETL process of work is associated with three steps:

Extract: Extraction is the process of accessing and retrieving information sources.

Transformation refers to the process of converting raw data into a form that meets the requirements of a set of guidelines. Data modification is performed when data is moved to a staging area for temporary storage.

Transformed and loaded data can be searched and used to create combination visualizations as well as prediction (at the predictive stage of descriptive analytics) (AltexSoft, 2019).

Data Infrastructure Implementation: Much of the structured data gathered is unified, stored, and analysed in the data warehouse. There are different departments with data of distinct characteristics that are typical in businesses (AltexSoft, 2019). In the present case, the data warehouse could be decomposed into smaller units, known as data marts, to facilitate and accelerate data querying (AltexSoft, 2019). AltexSoft (2019) agrees that data warehouses differ from operational databases because they focus on merging data from multiple time intervals rather than updating vast amounts of data as soon as possible.

Developing a Data Model: The data analytics would examine a data warehouse to determine whether it contains the information needed to address a particular problem. They start to collect more information in case of inadequate information. Once the data is ready for use in the machine learning process, the experts begin training the models (AltexSoft, 2019). Data with mapping variables (also referred to as labels) provides an opportunity to predict known characteristics from new information; this type of approach in training a model is called supervised learning (AltexSoft, 2019). The latter would lead to the eventual development of a so-called machine learning algorithm to identify similarities in revised data by learning to detect patterns in the dataset that align the features of the data frames with the correct answer (AltexSoft, 2019).

Unsupervised learning works with training data that is not answer-mapped (AltexSoft, 2019). Thanks to the implementation of HR use cases, it will be possible to cluster employees based on similarities or differences in their data, revealing implicit patterns and identifying the standards that mark good or poor performers (AltexSoft, 2019). HR Analytics Access: The key objective of this step is to create a front-end system, which allows corporate clients with no expertise in the fields of big data, analytics, and data science to create and adjust data and generate custom reports using a drag-and-drop interface: necessary elements and combine datasets to analyse (AltexSoft, 2019). Such desirables can be illustrated as follows: There are two types of dashboards: "static and dynamic (interactive)". The latter is data-driven and allows consumers to work with the data visualization. It offers many chart options and the ability to create new reports and present dashboards in minutes. The alternative is uploading custom graphics.

Libraries that provide data visualization are always there to speed development in place (AltexSoft, 2019). HR analysis is transforming the way HR contributes to the business. It may be due to a less effective selection process in the past and the lack of supporting data. Nevertheless, the right analytics will never cease to change the work front (Kumar, 2020). A significant part of knowledge is used to generate insights and, subsequently, transform them into stronger strategic knowledge about the functions in operation (Kumar, 2020).

To maintain strategic and operational power, human resource analysis therefore aims to develop new human-directed perceptions and ideas (Kumar, 2020). The idea behind time unit analytics rests on awareness and is a serious investment rather than speculation or imagination (Kumar, 2020).

When the HR activities are more automated, the organization saves time and energy. HR departments have now enhanced their employment and performance processes since the introduction of HR analytics (Bobriakov, 2020). Moreover, people-oriented, rather than profit-oriented, plans and strategies accelerate the retention process (Bobriakov, 2020). The human resource approach is practical and improves workforce performance through advanced analytics. The workers will perform tasks quickly and much more productively (Bobriakov, 2020).

The higher the employees' overall knowledge level, the happier and more satisfied they are with their jobs. The best talent development is internal training and development (Kumar, 2020). Most employers and companies have acknowledged the importance of analytics to HRM; however, further research and development can still be conducted on the role of analytics across the various spheres of HRM (Kumar, 2020).

HR analytics will need the support of HR managers in executing their HR duties such as facilitating demand and supply forecasting of staff positions, focusing on specific employment tests in a bid to cater to the characteristics of applicants, assessing the training and development requirements of employees, rolling out performance pay, and ensuring that they have the appropriate performance data to estimate bonuses and increase workplace restraint (Mohammed, 2020).

4.3 Human Resource Analytics Development

Technology is increasingly needed to manage personnel data, and demand is rising significantly. Information technology professionals have been allowed to help HR departments meet their requirements by developing an appropriate solution comprising both software and hardware. As technology advanced, it became clear that businesses could meet their technical needs more easily and achieve results at an affordable price. These technological advancements made it easier for the HR departments to familiarize themselves with the Information System and use it in their daily operations. A Human Resource Information System (HRIS) has emerged with advanced capabilities, encompassing reporting and record-keeping, as well as analytical capabilities used for decision-making.

A human resource information system has helped with payroll, performance management, benefits management, time management, recruitment, and compensation management. SAP, IBM, Oracle, and Microsoft are examples of business intelligence suppliers that offer data analytics services (Mohammed, 2020). Data scientists use Python for statistical data visualization and analysis (Mohammed, 2020). A programme for data analysis and visualization that can work with large datasets is RStudio (Mohammed, 2020). For a long time, Microsoft Excel has been a widely used tool for data collection, analysis, and transformation (Mohammed, 2020). Software from Microsoft Power Business Intelligence may collect data from multiple sources and facilitate data aggregation, analysis, and visualization (Mohammed, 2020).

As the business world has evolved, it has generated monumental amounts of data across all business processes (Gaur, 2021). Significantly greater use of Analytics affects all areas of the business, whereas HR use decreased; thus, the need to develop HR analytics arises due to technical issues in collecting and interpreting HR data (Gaur, 2021). HR analytics has grown since then, but now the data is developing quickly, even though it is still in its infancy (Gaur, 2021). Therefore, interest in HR analytics research has recently emerged, with an orientation towards its application as a decision-support tool, the development of skills, and the acquisition of HR analytics competence (Gaur, 2021).

4.4 Analysis of Data Quality Challenges Faced by HR Managers in Implementing HRA

The practical application of HRA enhances employee performance by helping them identify their strengths and development areas. An example is predictive analytics, which may help the HR group foresee performance issues and offer solutions before they occur, including targeted training (Gartner, 2023). HRA can be used to make performance appraisals more relevant, as it allows following up on current data across various metrics and fostering a culture of continuous improvement. According to McKinsey (2023), employees who receive consistent, data-driven, positive feedback tend to perform exceptionally well, as it aligns their personal targets with organizational outputs.

The insights produced by HRA have a significant effect on the retention rates. By examining employee turnover data, HR teams can identify trends and factors that drive disengagement or dissatisfaction (Dhankhar & Singh, 2022). HR managers ought to do this by initiating retention strategies, such as improved work–life balance packages or career development, when predictive models indicate employees are likely to leave. According to a SHRM (2023) study, turnover rates decrease in organizations that predict and reduce turnover with the help of HRA tools, as interventions enhance employee loyalty and commitment.

HRA will also help in employee career development. Organizations could use HRA tools to address skill deficits by designing individualized training programmes that match individual development requirements. The Top Employers Institute (2023) argues that when employees and employers align their career development, job satisfaction will increase and retention will be higher. When employees feel encouraged in their professional development, they are well adjusted to remain with the organization over time. To conclude, HRA plays a key role in achieving positive employee outcomes and high job satisfaction, engagement, performance, retention, and career development, all of which are success factors for the organization.

HRA also has a significant impact on job satisfaction and employee engagement, as it uses data-driven findings to resolve workforce issues. HRA helps organizations to reveal the most important drivers of engagement by analysing employee surveys and feedback, as well as performance indicators. Such a specific action will assist organizations in developing strategies to improve satisfaction and align employees with organizational objectives, thereby increasing productivity and morale (Biswas, 2023; CentraHub CRM, 2023). HRA also contributes to individual employee development by categorizing skills gaps and customizing training programmes. Such a fit between employee desires and professional development will lead to higher levels of satisfaction and retention (Top Employers Institute, 2023). HRA tools also generate predictive analytics that enable the HR team to proactively respond to the risk of disengagement or turnover by providing timely interventions (Avrahami et al., 2022).

HRA-supported real-time feedback systems complement the workplace communication, recognition, and continuous improvement. Employees also feel appreciated, making them even more committed and engaged (CentraHub CRM, 2023). In a broader sense, HRA helps maximize employees' experience by eliminating inefficiencies and balancing work and life, thereby increasing job satisfaction further (Top Employers Institute, 2023). To be effectively implemented, an organizational culture that accepts data-based decision-making and invests in relevant analytical tools and cross-functional knowledge will be necessary. Therefore, HRA is becoming a critical tool for increasing employee engagement, satisfaction, and an organization's success.

HRA is essential for improving employee performance by providing evidence-based information on workforce dynamics. HRA enables organizations to identify performance gaps and introduce personalized interventions by analysing performance metrics, sentiment surveys, and engagement data. To illustrate, predictive analytics can anticipate potential areas of performance concerns and inform personalized training and development programmes to enhance productivity (Avrahami et al., 2022). HRA also enables consistent performance tracking and provides real-time feedback for employees and managers. This is an iterative strategy that ensures individual goals and organizational objectives are aligned and, in turn, enables sustained improvement (McKinsey, 2023). Through such insights, organizations can maximize workforce efforts and drive the overall success of the business.

Human Resource Analytics helps ensure fair, open-minded assessment procedures, eliminating bias and increasing organizational trust. This will introduce a culture of fairness and accountability, which is vital in the induction of high performance. The effective use of HRA enhances decision-making quality, and such HR teams are prepared to allocate resources effectively and focus on strategic projects that increase employee productivity (SHRM, 2023). The challenges organizations face in achieving successful integration include data quality, technology adoption, and gaps in analytical skills. Training HR professionals in the use of HR analytics and data governance is a critical measure to maximize the potential of HRA in improving employee performance (Zebua et al., 2024). By integrating the best analytics and strategic HR practices, organizations can foster a culture of continuous improvement and individual performance aligned with corporate objectives.

4.5 Human Resource Analytics and Engagement Level

Data analytics provides an effective way to better understand employees and their level of engagement (Kumar, 2020). It can change the time and area of the HR function (Kumar, 2020). It could be used to manage attrition, define performance criteria, enhance employee relationships, and reduce prejudice during hiring (Kumar, 2020). Most organizations assume that analytics are essential and influence how they manage their operations (Kumar, 2020). One of the HR managers at Deloitte asserts that HR branches are evaluating the effectiveness of the right analytics to understand risks and make timely decisions supported by data collected over the years (Kumar, 2020).

4.6 Analytical Skills of HR Managers and Impact of Human Resource Analytics

The effectiveness of Human Resource Analytics (HRA) in organizations depends on HR managers' analytical skills. The emergence of HRA as a critical factor in data-driven decision-making poses a significant challenge for HR professionals, who need a sound understanding of analytical tools and methods to extract actionable insights from complex workforce data. In the absence of this skill set, the most sophisticated HRA systems cannot go far in facilitating organizational change.

Data Interpretation and Decision-Making: HR managers with strong analytical skills can interpret data more efficiently and ensure that insights align with the organization's objectives. Using data on employee performance, turnover rates, and engagement, they can identify patterns, potential problems, and ways to improve them (McKinsey, 2023; Yahaya et al., 2022). This power to make decisions based on data is the core of improving HR operations and ensures that policies align with strategic priorities.

Predictive Analytics: Predictive analytics. Teaching analytical skills will allow HR managers to predict future trends, such as employee turnover or engagement rates. The foresight will enable HR to take proactive steps to mitigate risks, enhance retention, and design training programmes aligned with the emerging needs of the labour force (Gartner, 2023). The more HR managers get familiar with data modelling and analytics, the higher their forecasting and responsiveness.

The construction of Data-Driven Cultures: HR individuals who are analytical help create a culture of data in the organization. Their functions are necessary to encourage the application of HRA tools across departments and to ensure that data is used to drive continuous improvement. Such a change in the culture makes organizations efficient and productive (SHRM, 2023).

4.7 Riding Analytical Skill Gaps

Although analytical skills are important, a lack of technical knowledge in data analytics poses a challenge for many HR managers (Peero, 2023). To optimize the impact of HRA systems, organizations will have to invest in training HR specialists in analytics tools and methods.

HRA is a highly sensitive subject that raises ethical concerns about the utilization of data in organizations that intend to leverage it responsibly. The more HR departments use data-driven insights in decision-making, the more ethical issues arise, particularly around privacy, discrimination, and transparency.

The collection and use of personal employee information and data are among the most significant ethical issues in HRA. It can be awkward or offensive to the employees when their performance, behaviour, or even health information is utilized without their knowledge (SHRM, 2023). The law on privacy, including the General Data Protection Regulation (GDPR), emphasizes the need for stringent restrictions on the processing of personal data. However, these issues are urgent as the use of sensitive data grows with the ongoing development of HR systems (Gartner, 2023).

Predictive analytics-based HRA systems (particularly) can have the unintended effect of perpetuating bias or discrimination. The information gathered about the history of hiring, promotions, and performance might include instances of discrimination based on gender, ethnicity, or other factors (Malla et al., 2023). Unless algorithms are designed and monitored, there is a risk of reinforcing these biases and creating discriminatory trends in the hiring or promotion process. HR managers should ensure that algorithms are verified for fairness and do not perpetuate existing inequities (McKinsey, 2023).

The other ethical concern in the use of HRA is transparency. Employees and other stakeholders need to know how data is used to make decisions about their careers. Lack of transparency will make employees feel that decisions are made arbitrarily, which might undermine trust and morale (Intervue, 2023). Organizations are expected to make the communication regarding the impact of HRA data on organizational decisions unambiguous, along with an explanation of the necessity of using a particular set of metrics.

4.8 *HR Managers' Analytical Skills and the Effectiveness of Human Resource Analytics*

Wang et al. (2024) used a thorough analysis and a dynamic framework for further study to investigate the factors that contribute to successful HR analytics implementation. The application of human resources (HR) analytics is an area that is constantly changing to keep up with the rapid advancement of digital technology, the authors emphasized. However, there is a dearth of understanding on the proper deployment of HR analytics, despite its practical value.

Hence, the study offers a detailed assessment of the factors enabling the successful application of HR analytics in business and provides a comprehensive critique of the 89 peer-reviewed papers on HR analytics published over the last two decades. Further, the study guides HR practitioners and scholars by proposing an emerging paradigm for implementing HR analytics grounded in adaptive structuration theory. The study also concludes with a research agenda to promote further research on HR analytics.

4.9 *Ethical Concerns and the Use of Human Resource Analytics*

Ethics and data privacy are important to the reliability of the healthcare process, as Rakesh (2023) investigated. To improve customer care and healthcare research, the healthcare sector has been using data-driven technologies, which are in step with the rapidly changing digital environment today. Nevertheless, this transformation also presupposes that today it is more important to protect patients' confidentiality and to act in accordance with ethical principles. The researchers emphasize the paramount importance of adherence to data privacy and ethics in healthcare, and the necessity of guaranteeing patient trust and the validity of medical studies. To prevent unauthorized use of patient information by hacking, high security measures, such as encryption and access limitations, are to be introduced.

The key points of ethical patient data management are informed consent collection and transparent communication. As healthcare increasingly relies on Artificial Intelligence (AI) and large-scale data analytics, it must adopt privacy-preserving processes that weigh heavily on technological advancement and patient privacy alike. Not only encouraging patients to make choices based on their data privacy, but also providing them with data concerning their health, facilitating openness and integrity in the policies, and strengthening the foundations of trust in the healthcare sector. It is not only a legal requirement but also an ethical obligation to observe data privacy and ethics in the best interest of the patient, which is the most important, and to promote the growth of the healthcare sector, which is and has always been dynamic.

5 Recommendations

The successful implementation of HRA is based on a clear analytics strategy. This involves formulating a clear set of goals, determining Key Performance Indicators (KPIs), and aligning analytics projects with organizational goals. Organizations are encouraged to invest in robust data collection and integration tools that can combine data from diverse sources, such as employee surveys, performance measures, and HR databases (Gartner, 2023).

It is also essential to train HR professionals to think analytically and provide them with access to modern tools such as predictive analytics and machine learning, so that analytical work can be translated into actionable insights (Top Employers Institute, 2023). Also, organizations ought to emphasize ongoing assessment and refinement of their analytics infrastructure to remain agile in the face of evolving business requirements and technological innovations.

Essential cooperation between the HR and IT departments is key to the successful implementation of HRA systems. IT expertise will be needed to ensure data security, system scalability, and the smooth integration of the analytics platform with the existing HR management systems. The combined teams should strive to create a user-friendly interface that enables HR professionals to access, analyse, and act on data insights easily (SHRM, 2023).

Knowledge sharing and problem-solving are also achieved through the development of cross-functional teams. To illustrate, IT experts can overcome technical issues, whereas HR staff provide domain expertise in workforce dynamics. The partnership guarantees that HRA tools are technologically advanced and practical in HR practices.

One obstacle to implementing HRA is ethical considerations regarding employee privacy and the use of employee data. Companies should develop detailed ethical codes governing the collection, analysis, and sharing of employee data (McKinsey, 2023). Such guidelines must comply with local and international data protection regulations and include measures for obtaining informed consent from employees.

Trust is mainly achieved through transparency among employees (Salman et al., 2024). The use of data should be communicated, and analytics practices should be developed to have a positive impact on the organization and the workforce. It can also be followed by establishing an ethics committee to monitor HRA's activities and prevent abuse, encouraging responsibility.

6 Limitations of the Study

The findings may be irrelevant across organizations, as they have different organizational cultures, sizes, and industry-specific considerations. Reliance on the secondary data and a lack of empirical studies limit the sound validation of claims. Changes in HR technologies can render the study's findings obsolete, so it is important to update it regularly. It is hard to address ethical issues, as laws vary globally in terms of privacy and subjective interpretations of fairness. Their study also assumes that the analytical capabilities of HR professionals are at a baseline level and may vary significantly across regions and organizations.

7 Suggestions for Future Research

HR analytics remains a relatively novel concept, though the number of people interested in the field is increasing. Scholars, practitioners, researchers, and students can thus explore the possible influence of the digitization of HR and the increased amount of personnel data on HR decision-making and organizational outcomes. The current study demystifies HR analytics research by assessing its impact on organizational performance. In so doing, we expect that further studies will be conducted to better understand the impact of HR analytics on a business. Future studies on the adoption of HR analytics should focus on industry-specific issues, e.g., differences in data availability, regulations, and labour mobility across industries such as healthcare, manufacturing, and technology. These studies can reveal industry-related best practices and challenges. Longitudinal HR analytics studies can investigate how it influences overall effects on employee performance, satisfaction, and organizational outcomes over time. This kind of research would provide information on how sustained investment in data quality, analytical tools, and ethical practices can lead to long-term success, enabling strategic decision-making and flexibility in the face of changing business conditions.

8 Conclusion

This chapter shows that Human Resource Analytics (HRA) plays a significant beneficial role in enhancing job satisfaction, engagement, and employee performance at organizations. HRA has also been found to facilitate data-driven HR practices, which, in turn, enable the addressing of employees' needs and contribute to organizational development. However, there are critical challenges that prevent them from being fully realized. The problem of data quality poses a significant obstacle to the validity of insights derived from HRA, which explains why better data management strategies should be implemented. The ability of HR managers to analyse influences HRA initiatives directly, which is why skill improvement programmes in HR departments are suggested. The ethical aspects, such as transparency, privacy, and fairness, must also be considered, as they affect employees' trust in and acceptance of analytics. By strategically interfering with this, it will be possible to maximize the benefits of HRA and take a balanced course of action that will lead to both performance and trust in HR practices. The research findings indicate that data quality, skills, and ethical issues that limit the full potential of HRAs affect their effectiveness in increasing job satisfaction, engagement, and performance in organizations.

References

AltexSoft. (2019, October 3). How to successfully implement HR analytics and people analytics in a company. AltexSoft.

Antwi-Boasiako, K. (2023). Data protection and HR analytics in Ghana. *African HR Review, 2*(2), 21–41.

Avrahami, D., Pessach, D., Singer, G., & Chalutz Ben-Gal, H. (2022). A human resources analytics and machine-learning examination of turnover: Implications for theory and practice. *International Journal of Manpower, 43*(6), 1405–1424. https://doi.org/10.1108/IJM-12-2020-0548

Bassi, B. L. (2022). Raging debates in HR analytics. *Human Resource Management International Digest, 20*(2), 74–80. https://doi.org/10.1108/hrmid.2012.04420baa.010

Biswas, S. (2023). *Understanding the impact of HR analytics on workforce engagement*. Retrieved November 1st 2023, from CentraHub CRM Blog.

Bobriakov, I., (n.d.).January 16, 2020, Top data science use cases in HR. Data Science Central. Retrieved November 8, 2021, from https://www.datasciencecentral.com/profiles/blogs/top-data-science-usecases-in-hr.

CentraHub CRM. (2023). *HR analytics benefits: Enhancing employee satisfaction and productivity through data insights*. Retrieved November 1st 2023, from CentraHub CRM.

Chalutz Ben-Gal, H. (2019). An ROI-based review of HR analytics: Practical implementation tools. *Personnel Review, 48*(6), 1429–1448. https://doi.org/10.1108/PR-11-2017-0362

Chatterjee, S., Chaudhuri, R., Vrontis, D., & Siachou, E. (2022). Examining the dark side of human resource analytics: An empirical investigation using the privacy calculus approach. *International Journal of Manpower, 43*(1), 52–74. https://doi.org/10.1108/IJM-02-2021-0087

Davenport, T. H., & Prusak, L. (2000). *Working knowledge: How organizations manage what they know*. Harvard Business School Press.

Deloitte. (2023). Global human capital trends. Retrieved from https://www2.deloitte.com

Dhankhar, K., & Singh, A. (2022). Employees' adoption of HR analytics—A theoretical framework based on career construction theory. *Evidence-Based HRM*. https://doi.org/10.1108/EBHRM-02-2022-0053

Dunmade, E. O., Ajayi, M. A., & Jimoh, A. L. (2023). Globalization and its impact on human resource management practices in Pz Cussons Nigeria plc., Lagos. The journal of business. *Studies, 7*(1). https://www.fcm.esn.ac.lk/jbs/vol_07-1.html

Fadamiro, J., & Adeoye, O. (2019). *Analytical skills in HRM: A strategic imperative*. Rafia Publications.

Falletta, S. V., & Combs, W. L. (2021). The HR analytics cycle: A seven-step processfor building evidence-based and ethical HR analytics capabilities. *Journal of Work Applied Management, 13*(1), 51–68. https://doi.org/10.1108/jwam-03-2020-0020

Forbes. (2023). Artificial intelligence and HR analytics: Transforming the future of work.

Gallup. (2022). *The relationship between employee engagement and organizational performance*. Gallup Research Reports.

Gartner. (2023). *Addressing privacy and data quality in HR analytics*. Retrieved November 1st 2024 from www.google.org

Gaur, D. M. (2021, September 14). HR analytics in business: Role, opportunities,and challenges of using it. Academia.edu. Retrieved November 8, 2021, from https://www.academia.edu/52387755/HR_analytics_in_Business_Role_Opportunities_and_Challenges_of_Using_It.

Haddaway, N. R., Page, M. J., Pritchard, C. C., & McGuinness, L. A. (2022). PRISMA2020: An R package and shiny app for producing PRISMA 2020-compliant flow diagrams, with interactivity for optimised digital transparency and. *Open Synthesis Campbell Systematic Reviews, 18*, e1230. https://doi.org/10.1002/cl2.1230

Intervue. (2023). *5 key challenges big data pose to HR*. Retrieved 22nd August, 2024 from https://www.intervue.io

Jimoh, A. L. (2025). Talent management practice in the contemporaneity organization. *A Systematic Literature Review. Economic Insights—Trends and Challenges, 14*(1), 59–72. https://doi.org/10.51865/EITC.2025.01.04

Jimoh, A. L., Dunmade, E. O., & Salman, A. (2025). Green talent management and job performance: The significance of open innovation. *Journal of Intellectual Capital*, 1–18. https://doi.org/10.1108/JIC-03-2025-0069

Juneja, P. (n.d.). Database concepts and its application in HRIS. MSG Management Study Guide. Retrieved November 8, 2021, from https://www.managementstudyguide.com/database-concepts-and-itsapplication-in-hris.htm.

Juneja, P. (2020). Human resource analytics and data-driven decision making in organizations. *International Journal of Human Resource Management and Analytics, 2*(2), 115–123.

Kumar, M. B. (2020, September 25). 5 killer applications of data science in HR analytics - 360digitmg. 360digitmg.com. Retrieved November 8, 2021, from https://360digitmg.com/applications-of-data-science-in-hr-analytics.

Kumar, T. H., Kumari, Y. V. N., Rao, A. N., Kumari, C. L. M., Janaki, G., Lakshmi, P. A., & Krishna, J. S. (2022). HR analytics as a moderating role in effective employee engagement and its outcomes. *NeuroQuantology, 20*(8), 4913–4920.

Madhani, P. M. (2022). Enhancing HR performance with the application of Blockchain. *The Journal of Total Rewards, 31*(2), 36–48.

Madhani, P. M. (2023). Human resources analytics: Leveraging human resources for enhancing business performance. *SAGE Publications, 55*(1), 31–45.

Malla, J., Hemalatha, T., & Naidu, S. T. (2023). Leveraging HR analytics for data-driven decision making: A comprehensive review. *IJFANS International Journal of Food and Nutritional Sciences, 11*(10), 1774–1784.

Margherita, A. (2022). Human resources analytics: A systematization of research topics and directions for future research. *Human Resource Management Review, 32*(2), 100795. https://doi.org/10.1016/j.hrmr.2020.100795

Marler, J. H., & Boudreau, J. W. (2017). An evidence-based review of HR analytics and implications for HR practice. *Journal of Human Resource Management, 28*(2), 45–62.

McKinsey. (2023). *Harnessing the power of HR analytics to improve employee performance.* Retrieved 28 August 2024 from https://www.mckinsey.com

Mhlanga, T. (2023). Challenges in HR digitalization in South Africa. *South African Journal of Business Analytics, 2*(4), 112–119.

Mohammed, D. A. Q. (2020, February 23). HR analytics: A modern tool in HR for predictive decision making. SSRN. Retrieved November 30, 2021, from https://papers.ssrn.com/sol3/papers.cfm?abstract_id=3525328.

Mondore, S., Douthitt, S., & Carson, M. (2011). Maximizing the impact and effectiveness of HR analytics to drive business outcomes: People and strategy. *A Peer Reviewed Open Access International Journal, 10*(18), 30–39.

Mureithi, P., Wanjiru, N., & Gichuhi, D. (2023). HR analytics infrastructure in Kenya. *East African Business Studies, 6*(1), 234–244.

Olowookere, A., & Adegbite, S. (2022). Human resource analytics in Nigeria: A review. *West African Management Review, 1*(1), 1–9.

Peero. (2023). HR analytics (2023). *Making data-driven decisions for a better workforce.* Retrieved from 11th March, 20224 from https://peero.app

Qamar, Y., & Samad, T. A. (2022). Human resource analytics: A review and bibliometric analysis. *Personnel Review, 51*(1), 251–283. https://doi.org/10.1108/PR-04-2020-0247

Rakesh, M. (2023). Ethics and data privacy: The backbone of trustworthy healthcare practices. *Socio-economic and Humanistic Aspects for Township and Industry, 1*(2), 232–236.

Rana, N., & Sharma, P. (2023). Exploring the impact of HCM systems on HR performance: The role of cloud technology. *Journal of Information Systems and HR Management, 40*(3), 203–214.

Rasmussen, T., & Ulrich, D. (2015). How HR analytics enhances workforce planning and performance. *Journal of Business Strategy, 36*(1), 21–29.

Salman, A., Olawale, Y. A., & Jimoh, A. L. (2024). Job burnout and employee commitment. *Farabi Journal of Social Sciences., 10*(2). https://doi.org/10.26577/FJSS2024.v10.i2.2

Society for Human Resource Management (SHRM). (2023). *Leveraging HR analytics for better employee outcomes*. Retrieved from https://www.shrm.org

Top Employers Institute. (2023). *The benefits of HR analytics for employee engagement and organizational success*. Available at. Accessed November 2023.

Union, E. (2018). General data protection regulation (GDPR). *Official Journal of the European Union.*

Vignoli, M., Mattarelli, E., & Mäkinen, S. J. (2018). Experimenting with innovation in innovative spaces. *CERN IdeaSquare Journal of Experimental Innovation, 2*(1), 1–2.

Wang, L., Zhou, Y., Sanders, K., Marler, J. H. & Zou, Y. (2024). Determinants of effective HR analytics Implementation: An In-Depth review and a dynamic framework for future research. *ELSEVIER: Journal of Business Research, 170* (2024), 114312. Retrieved https://doi.org/10.1016/j.jbusres.2023.114312.

Werkhoven, J. (2019). Building synergies with HR analytics capabilities. *Advanced Series in Management, 23*(2019), 141–160.

Yahaya, I. J., Jimoh, A. L., & Ameen, A. (2022). Effect of work-life balance on the performance of female employees of selected banks in Ilorin. *Unizik journal of entrepreneurship (ENTREP)., 3*(2). https://www.unizikentrepjournal.com/journalv3n2

Zebua, D. K., Santosa, T. A., & Putra, F. D. (2024). The role of HR analytics in enhancing organizational performance: A review literature. *Indonesia Journal of Engineering and Education Technology (IJEET), 2*(2), 362–368.

Ibrahim Olanrewaju Lawal is an accomplished Business Lecturer currently based at the GBS/University of Suffolk's Birmingham Campus in the United Kingdom (UK). With over a decade of experience in higher education across two countries, he specializes in Business, Human Resource Management, Organisation behaviour, and Management studies. He is recognized as a Fellow of the Higher Education Academy (FHEA), United Kingdom, and a Certified Management & Business Educator (CMBE), United Kingdom, reflecting his commitment to excellence in teaching and professional standards. He holds a PhD in Management from the University of KwaZulu-Natal in South Africa, as well as an MSc and BSc in Human Resources and Industrial Relations from Lagos State University, Nigeria. Before his current role in the UK, he served for more than ten years at Kwara State University (KWASU) in Nigeria, where he progressed to the position of Senior Lecturer and held various administrative leadership roles, including Coordinator for the Professional Degree Programme and Examination Officer.

Strategic Human Resource Analytics and Intelligent Data-Driven Systems: A Conclusion

Mufaro Dzingirai and Sulaiman Olusegun Atiku

Abstract The huge datasets derived from employee profiling have justified the exigent need for data-driven insights into the human resource decision-making process. With increasing demand for real-time data, the transformative power of data-driven intelligence has augmented the efficiency and effectiveness of Human Resource (HR) practitioners. These practitioners are utilizing a multiplicity of methodological tools in modelling big data generated from employee records in real time. It is impetus to mention that operational efficiency can be ensured by detecting errors using advanced statistical tools and research designs, supporting evidence-based decisions to achieve a durable competitive edge in the labor market. As such, this concluding chapter interrogates the drivers of the adoption of strategic human resource analytics and data-driven systems, challenges faced in strategic human resource analytics, in the contemporary digital economy, practical and policy implications, and the further research agenda. This chapter concludes that the strategic human resource analytics and data-driven systems are the heart and soul of evidence-based strategic decisions for sustainable value creation through the robust application of descriptive, diagnostic, predictive, and cognitive analytics.

Keywords People analytics · Artificial intelligence · Prescriptive analytics · Internet of things · Cognitive analytics · Machine learning · Strategic human resource management · Big data analytics

M. Dzingirai (✉) · S. O. Atiku
Harold Pupkewitz Graduate School of Business, Namibia University of Science and Technology, Windhoek, Namibia
e-mail: mdzingirai@nust.na

M. Dzingirai
Faculty of Business and Communications, INTI International University, Nilai, Malaysia

S. O. Atiku
Department of Economic and Business Sciences, Walter Sisulu University, Mthatha, South Africa
e-mail: satiku@nust.na

M. Dzingirai and S. O. Atiku (eds.), *Strategic Human Resource Analytics and Intelligent Data-Driven Systems*, https://doi.org/10.1007/978-981-92-2623-8_12

1 Introduction

The COVID-19 pandemic emerged as a powerful force behind the adoption of digital business across the globe (Shahiduzzaman, 2025). The strategic managers of both private and public sectors are continuously looking for a plethora of ways to optimize business operations in the current world that heavily depends on advanced technology. In the context of human resource management, executives are more concerned with the application of robust intelligent data-oriented systems with the aim of ensuring operational efficiency. This has led to a clear inclination towards human algorithmic management which permits the statistical behavioral patterns to be detected in real time. This is evident since the Internet of Things, smart technology, machine learning, and big data analytics are revolutionizing human resource management practices.

The Human Resource (HR) professionals are embracing descriptive analytics to understand and visualize trends related to employee demographics, labor retention rates, and time-to-hire. This can assist in monitoring recruitment effectively and feed insights into evidence-based human resource management decisions. Moreover, descriptive analytics of employee data and records can help HR professionals to monitor absenteeism and measure the return on investment. The HR professionals can also utilize predictive analytics with the use of employee data, statistical techniques, and machine learning to anticipate the future workforce trends like performance gaps, recruitment success, and employee turnover. The data-driven insights can help managers to acquire talent, forecast future headcount needs, and optimize employee performance. It is of utmost importance to note that cognitive analytics can help HR practitioners in applying natural language processing, machine learning, and artificial intelligence to synthesize complex and large volumes of employee data by imitating human reasoning (El-Garem, 2026). It appears that cognitive analytics can foster the provision of personalized employee services like training and reduce bias in targeted talent development.

In light of the above, Grigorescu et al. (2021) documented that the human–technology interaction demands new competencies and skills on the side of HR practitioners. In the same vein, Sakib et al. (2025) expressed that digital human resource management demands the transformation of the traditional models of managing human resources. In the current digital economy, it is not surprising that HR practitioners are considering advanced technologies as strategic assets for enhancing employee performance. To foster organizational agility and workforce responsiveness, innovation in recruitment is fundamentally necessary to align human resource management practices with the macro and micro environmental demands. As such, data-driven strategic decisions are also preferred by government ministries, whereby they apply artificial intelligence in labor economics (Mohammed, 2026).

Although the convergence of technology and human resource practices has been widely witnessed in developed countries, the underdeveloped digital ecosystems, lower technological maturity, weak institutional support, and nascent data governance frameworks are common in developing countries (Ristandi et al., 2026; Shet,

2025). It is quite disturbing since many developing countries are associated with demographic dividends, which means they must be at the forefront, championing the adoption of advanced technology because of a greater proportion of young people who are widely accepted as digital natives.

2 Drivers of Strategic Human Resource Analytics and Data-Driven Systems

This section of the concluding chapter interrogates the various drivers of the adoption of strategic human resource analytics and data-driven systems. Firstly, the changing workforce demographics pressurize HR professionals to adopt advanced human resource analytics. According to Zahra et al. (2025), the Gen Z is entering the labor market and expects changes in workplace policies as they prefer the adoption of technology. In tandem with the digital era, the Gen Z has high proficiency in using advanced digital technology at work place. This trend forces HR professionals to rethink the essence of technology to avoid potential conflicts with Gen Z. According to Surugiu et al. (2025), Gen Z's behaviors, communication styles, and expectations are shaped by social media, the Internet of Things, and emerging technologies as they are normally described as digital natives. The Gen Z are addicted to technology as they interact in digital social networks which forces HR professionals to reimagine the essence of technology adoption at the workplace (Benitez-Marquez et al., 2022).

Secondly, advancement in technology is another driver of strategic human resource analytics and data-driven systems in human resource management. In this context, according to Khatoon et al. (2025), the technology-driven evolution in human resource management has recently witnessed the adoption of data analytics, automation of repetitive tasks, artificial intelligence, and cloud computing that transformed the talent acquisition, development, and retention. The data-based strategic decision processes. The human resource information systems accelerated the adoption of advanced technologies in order to monitor, record, and document the recruitment, selection, training, evaluation, and retention of employees (Ali et al., 2023). Moreover, the previous studies established that electronic human resource practices affect employee productivity (Shiferaw & Birbirsa, 2025). Nonetheless, data privacy and security concerns are some of the challenges that are faced by organizations when embracing technological tools.

Thirdly, top management support is a key factor that supports the adoption of strategic human resource analytics in organization. The top management provides the financial resources needed for the acquisition of the advanced human resource information systems that are required for information visualization and data management (Chong et al., 2026). It is within this context that the top management support is crucial for stimulating a data-driven culture and strategic alignment (Alam et al., 2026). The strategic direction for descriptive, predictive, prescriptive, and cognitive analytics is of great importance for augmenting talent management performance outcomes.

Fourthly, the technological capabilities and competencies of the workforce can also influence the adoption of strategic human resource analytics and data-based intelligent systems. The technological skills of the HR professional must be assessed so as to deal with resistance to change. The organizations that give HR practitioners who are equipped with technological skills can find it very easy to implement advanced human resource information systems that support evidence-based decision-making in the organization.

3 Challenges Faced in Strategic Human Resource Analytics in the Digital Economy

After presenting the drivers for the adoption of advanced technology, this section is dedicated to the critical examination of the challenges faced in implementing strategic human resource analytics. The use of social media, wearable technology, microphones, email, cellphones, and motion sensors has been debated in terms of ethical and privacy concerns when collecting and analyzing employee data (Kellogg et al., 2020; Leonard & Contractor, 2018; McCartney & Fu, 2022). The advanced technologies like machine learning, artificial intelligence, Internet of Things, and data mining enable HR professionals to employees' behaviors, sentiments, and personal views. However, relational data and voice recording can be monitored, which violates employees' privacy (McCartney & Fu, 2022; Tursunbayeva et al., 2022).

The other challenge is the lack of Information Technology (IT) skills among the HR professionals. The digital transformation in human resource management has demanded a new skill set (MaCartney & Fu, 2022). The HR professionals are faced with a challenge of limited skills and competencies related to the implementation and use of human resource information systems. The specialized skill sets associated with strategic human resource analytics can lead to resistance to change. Moreover, the employees faced a challenge of fear of hacking of sensitive personal data via phishing, insider threats, social engineering, and third-party vendor risks.

The book is structured as follows: Chapter: "Strategic Human Resource Analytics and Intelligent Data-Driven Systems: An Introduction." This chapter conceptualizes and establishes the basis of integrating intelligent systems into SHRA. It demonstrates how this integration enhances HR effectiveness, supports strategic foresight, and repositions HR from an administrative support function to a proactive strategic partner. The emerging trends in SHRA are provided, highlighting its relevance for HR practitioners and showcasing the importance of the regulatory framework for intelligence SHRA and decision-making. With such an evidence-based perspective, the chapter contributes to scholarly and practical debates on how intelligent SHRA can be leveraged to drive workforce sustainability and improve operational efficiency and organizational outcomes in both public and private sectors.

Chapter: "The Role of Strategic Human Resources Analytics in Promoting Organizational Performance: A Case Study of State Universities in Zimbabwe." Using the mixed-method approach, this chapter examines the role of strategic human resources analytics in promoting organizational performance in three identified state universities. The research study revealed that Zimbabwean state institutions of higher learning lag in strategic human resource analytics adoption due to multiple challenges like financial restrictions, skill gaps, and cultural opposition. Nevertheless, participants acknowledge the possibility of improving productivity, equity, and important decision-making. A government-backed model, combined with training and low-cost technology, could drive SHRA integration. Future research should explore cost-effective analytics tools and change-management strategies for higher education. The findings contribute to the growing body of strategic human resource analytics and provide a unique contribution to theory, practice, and policy in higher education.

Chapter: "The Influence of AI Threats on Employee Behavior and Turnover Intentions." This chapter explores the relationship between the Threat of Artificial Intelligence (TAI) and employee turnover intention, emphasizing the roles of role ambiguity and knowledge hiding behavior. This study employs a quantitative approach using PLS-SEM to analyze how perceived Artificial Intelligence (AI) threats influence turnover intentions. The findings indicated that the Threat of AI (TAI) had a significant positive impact on Turnover Intention ($\beta = 0.289, p < 0.001$), Role Ambiguity ($\beta = 0.368, p < 0.001$), and Knowledge Hiding Behavior ($\beta = 0.670$, $p < 0.001$). Additionally, Role Ambiguity ($\beta = 0.119$, $p < 0.01$) and Knowledge Hiding Behavior ($\beta = 0.259$, $p < 0.01$) significantly increased turnover intention, while the mediating effect of role ambiguity (H4) was not supported. The implications of these findings highlight the need for organizations to foster transparent communication and supportive environments to reduce role ambiguity and prevent knowledge hiding.

Chapter: "Enhancing Employee Data Management in Uganda's Private Sector: Challenges, Benefits, and Pathways to Responsible Practices." This chapter combines a literature review with empirical evidence gathered via case studies, semi-structured interviews, and surveys, which is why it is a mixed-method study. Thematic analysis showed that although a few organizations have implemented organized data protection, some organizations have issues with compliance, technological constraints, and resource constraints. The results provide some of the major advantages of responsible data management, such as more trust between the employer and the employees, better legal and regulatory adherence, and more operational efficiency. Nevertheless, there are still major hindrances that include a lack of awareness, a poor application of data protection regulations, and the high cost of installing an effective data protection system. The chapter highlights the importance of well-defined policies, educating employees, and enhanced regulatory control to guarantee that the use of employee data in the Ugandan privacy sector is carried out in an ethical and efficient way.

Chapter: "Leveraging Personalized Employee Development Plans to Support Change Management Models in Academic Libraries in Zimbabwe." This chapter explores how personalized employee development plans assist academic libraries in Zimbabwe in implementing change management models. The study found that most academic libraries did not have custom employee development plans to support their change process. The study further stated that the potential value of custom employee development plans could support academic libraries in training their librarians with the skills needed to enable them to manage change. The study suggested that the Personalized employee development plans supported strategic planning, communication, and collaboration. The study concluded that the academic libraries in Zimbabwe would benefit from using custom employee development plans to support change management models. The study recommended that academic libraries should embrace custom staff development plans to enhance librarians' capacity for managing change and embed them into their change management component.

Chapter: "**Exploration of Artificial Intelligence as a Catalyst for Human Resource Management Effectiveness**." This research investigated the relationship between predictive performance analytics and performance management. It also examined the effect of the robotics automation process on the recruitment and selection process. Descriptive research design was adopted in this study using 148 workers of BeaconGate Limited as the study population. Research findings confirmed a correlation between the utilization of predictive performance analytics and performance management and likewise an influence of robotics process automation on the recruitment and selection process. This study offered insightful information about how to integrate AI technology into an organization's human resource management strategy for effectiveness.

Chapter: "**Ethics In Human Resource Analytics: Perspectives from Public and Private Organizations in Ghana**." This chapter examines the ethical ramifications of Human Resource Analytics (HRA) within Ghana's public and private sectors, focusing on justice, privacy, transparency, and ethical governance. Five major themes emerged in relation to problems, namely, algorithmic bias and fairness, data privacy and surveillance, employee autonomy and permission, profiling and behavioral trust, and organizational culture and ethical governance. The results showed that in Ghanaian businesses, the lack of digital infrastructure, low levels of data literacy, and weak ethical frameworks made discrimination and privacy violations more likely. Cultural hierarchy and regulatory gaps often make it harder to be open and responsible when using analytics. However, some companies that are ahead of the curve, especially in banking and telecommunications, have started to include fairness audits, ethical review processes, and compliance with Ghana's Data Protection Act (2012, Act 843). The study contributes to the evolving African discourse on ethical analytics by illustrating that ethical governance, trust, and cultural alignment are essential for responsible data practices.

Chapter: " Adaptive Talent Acquisition Models for Navigating Change in a Dynamic Environment." This chapter examines the transformation of talent acquisition models for contributing to organizational success in a dynamic environment. In a country like India, labor markets are dynamic, and skill set requirements are also ever-changing. Hence, the initial part of the study examines the features of various types of talent acquisition models followed in India and also at the global level. As a progress study exhibits the talent acquisition process with special emphasis on employer branding and Employee Value Proposition (EVP). The study also delves into the impact of technology and automation on talent acquisition practices. The last part of the study examines the challenges of AI in HR practices, especially in the context of the Volatile, Uncertain, Complex, and Ambiguous (VUCA) world and concludes by sketching the road ahead for talent acquisition models, which leverage data-driven insights, technology, and agile recruitment strategies to build resilient talent pipelines that align with evolving business needs.

Chapter: " Cybersecurity Vulnerability and Organizational Performance: A Systematic Literature Review." This chapter researches the cybersecurity weaknesses and how this weakness affects the organizational efficiency. To examine the associations between cybersecurity threats, such as those of hacktivism, cybercrime, and cyber spying and the key performance indicators, such as reputational damage, operational effectiveness, and financial stability, a systematic literature review method was used. Data were collected via EBSCOhost, Elsevier, ProQuest, Wiley-20, SpringerLink, and Google Scholar and were reviewed and analyzed through Systematic Reviews and Meta-analyses (PRISMA). As technology and information systems expand globally, businesses must stay ahead of the curve by enhancing security measures and proactively resolving authentication and authorization vulnerabilities that could jeopardize confidentiality and availability. Research shows that hacktivism, cybercrime, and cyber espionage harm organizational performance. To combat these risks, companies must create backup systems to protect client and customer information from hostile threats.

Chapter: " Ethical Dilemmas of Generative AI and Large Language Models in Human Resource Management." The rapid technological advancements witnessed in the early twenty-first century are reforming the human resource management at a faster pace. However, there is scant evidence pertaining to the influence of generative AI and Large LLMs on human resource management. This chapter seeks to bridge this gap. This chapter aims to examine how generative AI and LLMs influence talent management in a data-oriented context. A desktop approach and a conceptual approach were used as the data collection technique. The results revealed that making use of generative AI and LLMs alters the human resource management within a data-oriented setting. On the other side, the results showed that risks and ethical dilemmas affect the use of generative AI and LLMs within the data-oriented workplace. These results are useful for human resource practitioners, policymakers and scholars.

Chapter: " Human Resource Analytics: Challenges and Prospects for Human Resource Effectiveness." Although Human Resource Analytics (HRA) is increasingly used and the question of how it can influence employee productivity remains unanswered. Human Resource Analytics is an enhancement of Human Resources practice that uses data analysis to advance HR practices, improve employee engagement, performance, and decision-making, and enhance organizational growth, efficiency, and strategic workforce planning. The current research aims to fill this gap by investigating the impact of HRA on job satisfaction, engagement, performance, and other aspects of the workplace. The study adopted the Systematic Reviews and Meta-analyses (PRISMA) to achieve smooth, superior, and improved comprehension of the examination of the gathered information and analysis. The results indicate that HRA is positively related to employee satisfaction, engagement, and overall performance, thereby enhancing HR's capacity to address workforce needs. Nevertheless, issues of low data quality, lack of analytical skills, and ethical concerns, especially regarding privacy and fairness, affect the effectiveness and credibility of HRA initiatives. To get the best out of HRA, the chapter suggests that organizations adopt effective data management systems, improve HR professionals' analytical capabilities through training, and develop ethical standards. Such measures will contribute to transparency, employee confidence, and the successful implementation of data-oriented HR practices.

Chapter: " **Strategic Human Resource Analytics and Intelligent Data-Driven Systems: A Conclusion**." This concluding chapter interrogates the drivers of the adoption of strategic human resource analytics and data-driven systems, challenges faced in strategic human resource analytics, in the contemporary digital economy, practical and policy implications, and the further research agenda. This chapter concludes that the strategic human resource analytics and data-driven systems are the heart and soul of evidence-based strategic decisions for sustainable value creation through the robust application of descriptive, diagnostic, predictive, and cognitive analytics.

4 Practical and Policy Implications

Based on the critical examination of the issues surrounding strategic human resource analytics, the following practical and policy implications are proposed:

The policymakers should create a conducive legal and operating environment by reducing taxes on the importation of technological products. This could allow the companies to reduce technological costs and then increase the technological adoption by companies. Moreover, policymakers should set aside a ring-fenced money for Research and Development. As such, the installation of Human Resource Information systems. In terms of organizations, it is hereby recommended that they invest large sums of money in advanced technologies such as machine learning, Large Language Models (LLMs), and artificial intelligence. This is of utmost importance in the context of the deployment of user-friendly dashboards for HR professionals as well as in the collection and dissemination of real-time employee data and information.

To effectively address data integrity and security challenges, proactive and agile organizations should implement robust employee data governance frameworks. These frameworks can reduce the negative impact of legal risk as they can eliminate legal liabilities associated with the failure to safeguard employee data in line with strict data standards within the context of collection, storage, analysis, and sharing of sensitive employee datasets. Robust data governance allows HR professionals to secure and protect employee datasets while ensuring effective data-oriented strategic human resource decisions.

Moreover, organizations should align human resource analytics with the business strategy. The executives of organizations can connect HR metrics like performance appraisal indicators and turnover measures to strategic goals of the organization like productivity and revenue growth as well as cost minimization. Lastly, the organizations should also develop HR team's capabilities in data analytics tools like machine learning, Large Language Models (LLMs), and artificial intelligence. This can cover key issues of capabilities from data collection to data sharing using dashboards. As such, HR professionals would have the necessary technological skills that foster effective adoption of ethical, security, and privacy-oriented human resource practices in data-driven talent management systems.

5 Further Research Agenda

This book focused on strategic human resource analytics and intelligent data-driven systems which opened avenues for further research as follows:

- **Monitoring and Evaluation:** The research work can investigate the methods used to systematically collect data and evaluate the contribution of strategic human resource analytics and intelligence data-driven systems to return on investment. The comparative studies can add value in assessing the effect of HR analytics on increased productivity.
- **Longitudinal Human Resource Analytics studies:** As many previous studies have focused on cross-sectional data, it is prudent for future researchers to focus on longitudinal studies to predict human resource outcomes and analyze workforce trends. The prolonged studies could establish causality and give HR professionals insights into the "why" behind workforce trends.
- **Algorithmic Fairness:** The studies focusing on implementing and auditing AI systems in human resource management are welcomed in order to evaluate biased performance appraisal, hiring, and recruitment. The future research can be specifically dedicated to distributive fairness, procedural fairness, and data fairness within the context of AI-driven talent management.

6 Conclusion

This book chapter has delved deeply into the core tenets of the book by synthesizing various issues from all the chapters of this book. With increasing demand for real-time data, the transformative power of data-driven intelligence has augmented the efficiency and effectiveness of Human Resource (HR) practitioners. Although the intersectionality of data analytics and human resources has been hailed for robust evidence-driven strategic decisions in recruitment, development of personnel, performance appraisal, and employee retention, implementing strategic human resource analytics and intelligent data-driven systems is often associated with challenges like resistance to change, ethical and privacy concerns, and security threats. In light of evidence-based best practices, human resource information systems are crucial for better decision-making processes. This chapter provided the practical and policy recommendations as well as the future research directions.

References

Alam, S., Dong, Z., Kularatne, I., & Rashid, M. S. (2026). Exploring approaches to overcome challenges in adopting human resource analytics through stakeholder engagement. *Management Review Quarterly, 76*(1), 729–787.

Ali, O., Krsteska, K., Said, D., & Momin, M. (2023). Advanced technologies enabled human resources functions: Benefits, challenges, and functionalities: A systematic review. *Cogent Business & Management, 10*(2), 2216430.

Benítez-Márquez, M. D., Sánchez-Teba, E. M., Bermúdez-González, G., & Núñez-Rydman, E. S. (2022). Generation Z within the workforce and in the workplace: A bibliometric analysis. *Frontiers in Psychology, 12*, 736820.

Chong, L. Y. Q., Lim, T. S., & Nipo, D. T. (2026). Human resource analytics adoption and its impact on talent management and operational performance: Evidence from Malaysian firms. *Journal of Modelling in Management*, 1–20.

El Garem, R. A. (2026). HR digital transformation: Enhancing human resource management through technology. *Future Business Journal, 12*(1), 5.

Grigorescu, A., Pelinescu, E., Ion, A. E., & Dutcas, M. F. (2021). Human capital in digital economy: An empirical analysis of central and eastern European countries from the European Union. *Sustainability, 13*(4), 2020.

Kellogg, K. C., Valentine, M. A., & Christin, A. (2020). Algorithms at work: The new contested terrain of control. *Academy of Management Annals, 14*(1), 366–410.

Khatoon, U. T., Babgi, M., Hadi, N. T., Mir, R. N., & Velidandi, A. (2025). Technology-driven change in human resource management: Reshaping talent management and organizational design. *Administrative Sciences, 15*(11), 452.

Leonardi, P., & Contractor, N. (2018). Better people analytics. *Harvard Business Review, 96*(6), 70–81.

McCartney, S., & Fu, N. (2022). Promise versus reality: A systematic review of the ongoing debates in people analytics. *Journal of Organizational Effectiveness: People and Performance, 9*(2), 281–311.

Mohammed, S. M. (2026). Workforce transformation: The role of artificial intelligence in the economics of government human resources in service ministries in KSA. *International Journal of Productivity and Performance Management, 75*(2), 672–691.

Ristandi, D. K., Maarif, M. S., Affandi, J., & Sukmawati, A. (2026). Ecosystem-driven adoption of human resource analytics: Institutional mechanisms and perceived organizational performance in Indonesia. *Cogent Business & Management, 13*(1), 2642549.

Sakib, M. N., Ullah, M. S., & Rahman, M. M. (2025). Mapping the evolution of digital human resource management: A systematic review and bibliometric analysis. *Future Business Journal, 11*(1), 154.

Shahiduzzaman, M. (2025). Digital maturity in transforming human resource management in the post-COVID era: A thematic analysis. *Administrative Sciences, 15*(2), 51.

Shet, S. V. (2025). Developing human resource analytics capability architecture in organizations. *Personnel Review, 54*(9), 2286–2310.

Shiferaw, R. M., & Birbirsa, Z. A. (2025). Digital technology and human resource practices: A systematic literature review. *Heliyon, 11*(2).

Surugiu, C., Surugiu, M. R., Grădinaru, C., & Grigore, A. M. (2025). Factors motivating generation Z in the workplace: Managerial challenges and insights. *Administrative Sciences, 15*(1), 29.

Tursunbayeva, A., Pagliari, C., Di Lauro, S., & Antonelli, G. (2022). The ethics of people analytics: Risks, opportunities and recommendations. *Personnel Review, 51*(3), 900–921.
Zahra, Y., Handoyo, S., & Fajrianthi, F. (2025). A comprehensive overview of generation Z in the workplace: Insights from a scoping review. *SA Journal of Industrial Psychology, 51*, 2263.

Mufaro Dzingirai is a Senior Lecturer at Harold Pupkewitz Graduate School of Business, Namibia University of Science and Technology, Namibia and Research Fellow at INTI International University, Malaysia, as well as a Consultant at the Econowrite Institute for Policy Innovation & Development (EIPID), Australia.

Sulaiman Olusegun Atiku is a Director of Research and HRM Professor at Harold Pupkewitz Graduate School of Business, Namibia University of Science and Technology, Namibia and Adjunct Professor in the Department of Economic and Business Sciences, Walter Sisulu University, South Africa.

The manufacturer's authorised representative in the EU is Springer Nature Customer Service Centre GmbH, Europaplatz 3, 69115 Heidelberg, Germany. If you have any concerns regarding our products, please contact ProductSafety@springernature.com

Printed and bound by CPI Group (UK) Ltd, Croydon, CR0 4YY
21/09/2026
02219989-0011